2nd International Symposium on Optics and Its Applications (OPTICS-2014)

AA002731

Journal of Physics: Conference Series Volume 672

Yerevan and Ashtarak, Armenia
1 – 5 September 2014

Editors:

Maria L. Calvo
Irina N. Dolganova
Narine Gevorgyan
Angela Guzman
Aram Papoyan
Hayk Sarkisyan
Stanislav Yurchenko

ISBN: 978-1-5108-1963-4
ISSN: 1742-6588

Printed from e-media with permission by:

Curran Associates, Inc.
57 Morehouse Lane
Red Hook, NY 12571

Some format issues inherent in the e-media version may also appear in this print version.

Printed by Curran Associates, Inc. (2016)

Institute of Physics
Dirac House, Temple Back
Bristol BS1 6BE UK

Phone: 44 1 17 929 7481
Fax: 44 1 17 920 0979

techtracking@iop.org

Additional copies of this publication are available from:

Curran Associates, Inc.
57 Morehouse Lane
Red Hook, NY 12571 USA
Phone: 845-758-0400
Fax: 845-758-2634
Email: curran@proceedings.com
Web: www.proceedings.com

Table of contents

Volume 672

2nd International Symposium "Optics and its Applications"
1–5 September 2014, Yerevan-Ashtarak, Armenia

Accepted papers received: 27 November 2015
Published online: 20 January 2016

Journal of Physics: Conference Series **672** (2016) 011001 doi:10.1088/1742-6596/672/1/011001

The Abdus Salam International Centre for Theoretical Physics 50th Anniversary 1964–2014

2nd International Symposium "Optics and its Applications" Yerevan-Ashtarak, Armenia, September 1-5, 2014 Preface

The ICTP smr2633: 2nd International Symposium "Optics and its Applications" (OPTICS-2014) http://indico.ictp.it/event/a13253/ was held in Yerevan and Ashtarak, Armenia, on 1–5 September 2014. The Symposium was organized by the Abdus Salam International Center for Theoretical Physics (ICTP) with the collaboration of the SPIE Armenian Student Chapter, the Armenian TC of ICO, the Russian-Armenian University (RAU), the Institute for Physical Research of the National Academy of Sciences of Armenia (IPR of NAS), the Greek-Armenian industrial company LT-Pyrkal, and the Yerevan State University (YSU). The Symposium was co-organized by the BMSTU SPIE & OSA student chapters. The International Symposium OPTICS-2014 was dedicated to the 50th anniversary of the Abdus Salam International Center for Theoretical Physics. This symposium "Optics and its Applications" was the First Official ICTP Scientific Event in Armenia.

The presentations at OPTICS-2014 were centered on these topics: optical properties of nanostructures; quantum optics & information; singular optics and its applications; laser spectroscopy; strong field optics; nonlinear & ultrafast optics; photonics & fiber optics; optics of liquid crystals; and mathematical methods in optics.

OPTICS-2014 was attended by 146 scientists from 29 countries of five continents, namely: Armenia, Australia, Belgium, Cuba, Denmark, Egypt, France, Georgia, Germany, Greece, India, Iran, Italy, Japan, Latvia, Lithuania, Mexico, Moldova, Poland, Russia, Saudi Arabia, South Africa, Switzerland, Turkey, Ukraine, United Kingdom, USA, Uzbekistan, and Venezuela. The goal of this symposium was to bring together experienced and young scientists from various countries working on optics, and to provide a perfect setting for discussions of the most recent developments in that area. The symposium provided a good opportunity for contact and information exchange between experts and students.

The present volume of Journal of Physics: Conference Series includes proceedings of the symposium covering various aspects of modern problems in optics.

We would like to express our sincere gratitude to Joseph Niemela and Ave Lusenti for their guidance and invaluable help in preparing and performing this symposium. We would like to thank Gagik Buniatyan for his help and support in organizing OPTICS-2014, as well as all staff members of LT-Pyrkal who contributed to making the symposium happen.

We gratefully acknowledge support from: the Optical Society of America (OSA), the International Commission for Optics (ICO), the International Society for Optics and Photonics (SPIE), the RAU, LT-Pyrkal, and the Laboratory of Terahertz Technology of the Bauman Moscow State Technical University (BMSTU).

The Editors,
Maria L Calvo *Universidad Complutense de Madrid, Spain*
Irina N Dolganova *Bauman Moscow State Technical University, Russia*
Narine Gevorgyan *Russian-Armenian (Slavonic) University, Armenia; Abdus Salam International Center for Theoretical Physics, Italy*
Angela Guzman *CREOL, USA*
Aram Papoyan *Institute for Physical Research of NAS, Armenia*
Hayk Sarkisyan *Russian-Armenian (Slavonic) University, Armenia; Yerevan State University, Armenia*
Stanislav Yurchenko *Bauman Moscow State Technical University, Russia*

Journal of Physics: Conference Series **672** (2016) 011002 doi:10.1088/1742-6596/672/1/011002

2nd International Symposium "Optics and its Applications" Yerevan-Ashtarak, Armenia, September 1-5, 2014

ORGANIZERS

Directors:

Gagik Buniatyan (*LT-PYRKAL, Armenia*)
Maria L Calvo (*Universidad Complutense de Madrid, Spain*)
Narine Gevorgyan (*Russian-Armenian University, Armenia & ICTP, Italy*)
Angela M Guzman (*CREOL, USA*)
Joseph Niemela (*ICTP, Italy*)
Aram Papoyan (*Institute for Physical Research of NAS, Armenia*)
Hayk Sarkisyan (*Russian-Armenian University, Armenia & Yerevan State University, Armenia*)

International Advisory and Program Committees

Chairs: David Blaschke (*University of Wroclaw, Poland & JINR Dubna, Russia*)
David Sarkisyan (*Institute for Physical Research of NAS, Armenia*)

Members:
Roman Alaverdyan (*YSU, Armenia*)
Levon Aslanyan (*YSU, Armenia*)
Marcis Auzinsh (*University of Latvia, Latvia*)
Stefano Bellucci (*Frascati National Laboratory INFN, Italy*)
Michael Berry (*University of Bristol, UK*)
Aranya Bhattacherjee (*University of Delhi, India*)
Konstantin Bliokh (*Advanced Science Institute RIKEN, Japan*)
Dmitry Budker (*University of California at Berkeley, USA*)
Vigen Chaltykyan (*IPR of NAS, Armenia*)
Miltcho Danailov (*Elettra, Italy*)
Carlos Duque (*Universidad de Antioquia, Colombia*)
Vladimir Gerdt (*JINR Dubna, Russia*)
Gayane Grigoryan (*IPR of NAS, Armenia*)
Rafik Hakobyan (*YSU, Armenia*)
Alexey Kavokin (*University of Southampton, UK*)
Eduard Kazaryan (*RAU, Armenia*)
Arsen Khvedelidze (*Institute of Mathematics, Georgia*)
Albert Kirakosyan (*YSU, Armenia*)
Radik Kostanyan (*IPR of NAS, Armenia*)
Takayuki Miyadera (*Kyoto University, Japan*)
Levon Mouradian (*YSU, Armenia*)
Atom Muradyan (*YSU, Armenia*)
Armen Nersessian (*YSU, Armenia*)
Avinash Pandey (*University of Allahabad, India*)
Luis Ponce (*CICATA IPN, Mexico & Universidad de Havana, Cuba*)
Hartmut Ruhl (*LMU, Germany*)

Marat Soskin (*Institute of Physics NASU, Ukraine*)
Sergue Vinitsky (*JINR Dubna, Russia*)
Gert von Bally (*Centre for Biomedical Optics and Photonics, Germany*)
Ruben Zadoyan (*Newport Technology & Applications Center, USA*)

Local Organizing Committee

Secretary: **Ave Lusenti** (ICTP, Italy)

Coordinator: **Paytsar Mantashyan** (*IPR of NAS, Armenia*)

Managers:
David Hayrapetyan (*RAU, Armenia*)
Artsrun Martirosyan (*IPR of NAS, Armenia*)

Members:
Melanya Grigoryan (*Integrator Ltd, Armenia*)
Astghik Kuzanyan (*IPR of NAS, Armenia*)
Lilit Mantashyan (*Guw-Re Ltd Educational Centre, Armenia*)

Symposium was sponsored by

Sponsored by OSA — Optical Society

International Commission for Optics / Commission Internationale d'Optique — International Commission for Optics

SPIE — International Society for Optics and Photonics

Russian-Armenian (Slavonic) University

LT-PYRKAL — LT-Pyrkal

TT — TeraTech Laboratory (BMSTU)

Armenian SPIE Student Chapter

The Abdus Salam
International Centre
for Theoretical Physics
50th Anniversary 1964-2014

Photo Gallery

The Greek-Armenian industrial company LT-Pyrkal (5 min before the symposium opening)

Registration, Exhibition and Poster Session

Awarded students for best Students Oral Presentations
From left to right: Irina Dolganova (award for 2nd place), Naira Grigoryan (award for 2nd place), Yousef Pourvais (award for 1st place), and Joseph Niemela (co-director of OPTICS-2014).

*Historic ancient city of Yerevan – **Erebuni** (782 BC)*

doi:10.1088/1742-6596/672/1/011003

William Rhodes, Joseph Niemela, Stefano Bellucci, Armen Nersessian, Narine Gevorgyan, and David Blaschke

Michael Berry, Hayk Sarkisyan and Eduard Kazaryan

Janos Bergou and Alexey Chizhov

Mark Dennis and Luis Ponce, SPIE Visiting lecturers, invited by Armenian SPIE Student Chapter

Jean-luc Duomont, SPIE Visiting lecturer, invited by Armenian SPIE Student Chapter

Gagik Buniatyan and Vladimir Manko

Journal of Physics: Conference Series **672** (2016) 011003 doi:10.1088/1742-6596/672/1/011003

Participants of OPTICS-2014 in front of the Russian-Armenian University.

Presidium of the National Academy of Sciences of RA. One of the parallel sessions was a memorial seminar dedicated to the 70th anniversary of academician A.N. Sissakian, the former director of JINR Dubna.

Journal of Physics: Conference Series **672** (2016) 011003 doi:10.1088/1742-6596/672/1/011003

Celebration of 50th Anniversary of ICTP

Peer review statement

All papers published in this volume of *Journal of Physics: Conference Series* have been peer reviewed through processes administered by the proceedings Editors. Reviews were conducted by expert referees to the professional and scientific standards expected of a proceedings journal published by IOP Publishing.

Journal of Physics: Conference Series **672** (2016) 012001 doi:10.1088/1742-6596/672/1/012001

Numerical simulations of radiation transfer in partially-ordered stratified media using Monte Carlo methods

Aleksandra S Neganova, Irina N Dolganova and Valeriy E Karasik
Bauman Moscow State Technical University, 2nd Baumanskaya str. 5, 105005, Moscow, Russia
E-mail: in.dolganova@gmail.com

Abstract. This work includes the research method of terahertz radiation scattering process in partially-ordered stratified media. This method was set on algorithm of computational electrodynamics and Monte Carlo simulations. The method allows to determine differential and integral cross sections and scattering phase functions. The algorithm of modeling radiation transfer in stratified media using Monte Carlo approach was created. Rigorous and Henyey-Grinstein phase functions were considered. Analysis of simulated results was developed. The technique for accounting single scattering was designed.

1. Introduction

Multiple scattering problem is essential for such fields as atmosphere remote sensing, spectroscopy, biochemistry and others. Different aspects of this problem have been considered in literature [1–5].

According to this work we consider one of the essential cases of multiply scattering: scattering in partially-ordered medium with group formation, when particles with dimensions of the order of wavelength are combined densely in groups, and these groups are randomly distributed in the scattering volume.

The radiation transfer theory is one of the widely used approximated methods to describe multiple scattering processes [6]. It operates with the radiation transfer equation (RTE). In order to solve RTE numerically we need to use the Monte-Carlo (MC) approach [7].This method allows the parameters of inhomogeneous scattering media to be varied easily and different experimental geometries to be simulated.

When medium parameters differ within the volume, averaging over the volume results in inaccuracy of MC simulation. In this case we must divide the volume into layers with different parameters. When every layer consists of structured group of particles, it is rather complicated to understand what is the influence of the scattering phase function on radiation transfer results. The Henny-Greenstein phase function [8] is suitable in the event that medium consists of randomly distributed single particles. However, it is essential to take into account the impact of the interference of scattered waves on the group scattering phase function. The primary aim of the work is to compare Monte Carlo simulation results of using rigorous and Henyey-Greenstein phase functions.

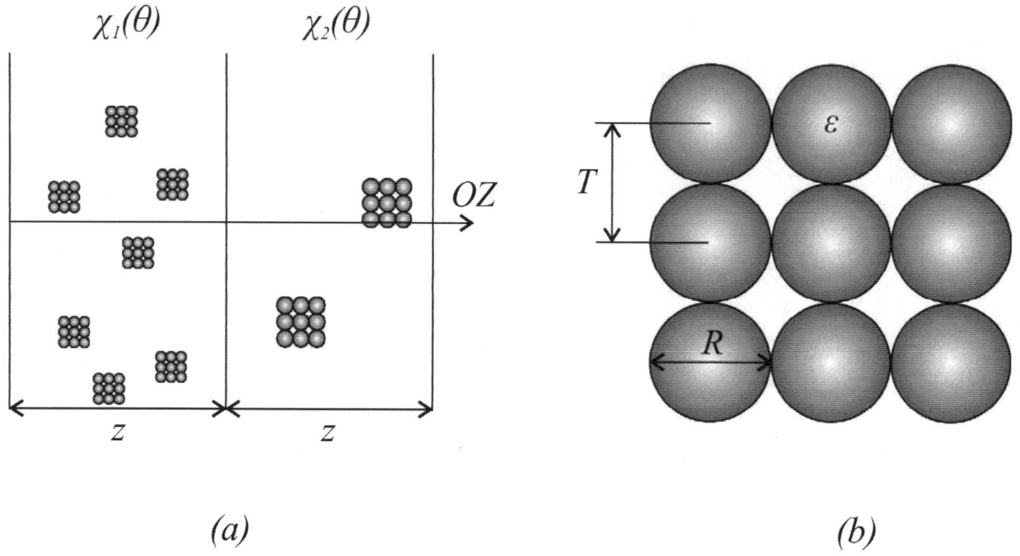

Figure 1. Scattering medium with 2 layers (a), model of particles in the group (2D)(b).

2. Scattering volume

We consider the arbitrary two-layered medium: each layer consists of randomly distributed groups of nine identical particles with circular cross section in YZ plane (figure 1b) and permittivity $\varepsilon_P = 1.96$. Dimensions of particles (table 1) in every layer are of the order of wavelength λ of TEM plane wave polarized radiation, incident in OZ direction (figure 1a).

Table 1. Geometric parameters of particle groups.

Layer	1	2
Particle radius, R/λ	1.00	1.50
Distance between particles centers, T/λ	2.50	3.75

3. Methods

3.1. Single scattering phase function

The single scattering phase function is the angular distribution of light scattered on object when it is illuminated. In order to find single scattering phase function $\chi(\theta)$ more accurately we use finite-difference time-domain method (FDTD) [9] of Maxwells equations numerical solution. As a result, we obtain the rigorous group single scattering phase function (figure 2a):

$$\chi_{rig}(\theta) = \frac{2\sigma(\theta)}{\int_0^{\pi} \sigma(\theta) sin(\theta) d\theta}, \tag{1}$$

where $\sigma(\theta)$ is the group scattering cross section.

Henyey-Greenstein phase function $\chi_{HG}(g,\theta)$ (figure 2b) [8] is calculated from the equation:

$$\chi_{HG}(g,\theta) = \frac{1-g^2}{(1+g^2-2gcos(\theta))^{3/2}}, \tag{2}$$

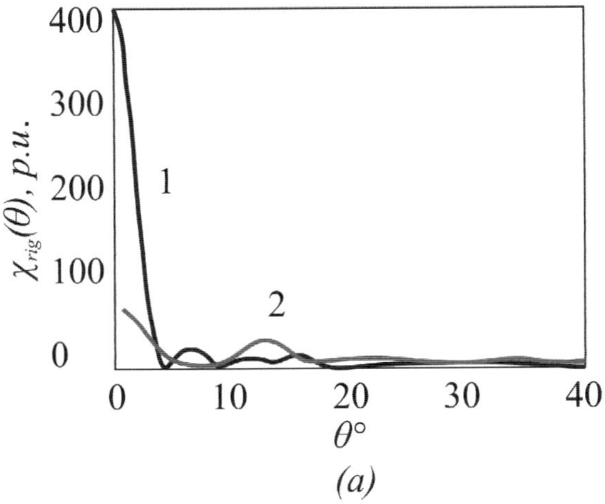

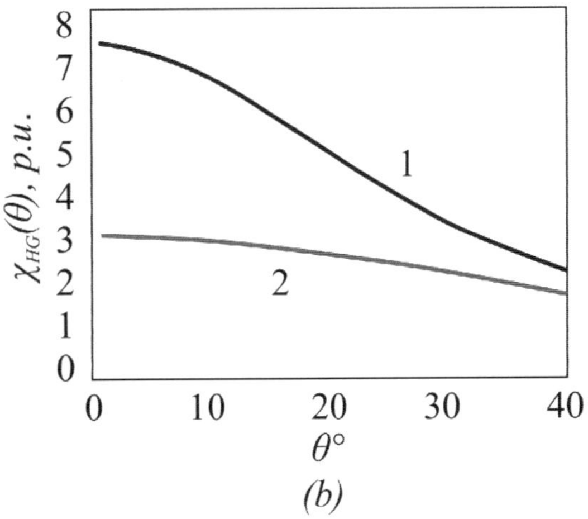

Figure 2. Rigorous scattering phase functions $\chi_{rig}(\theta)$ (a) and HG scattering phase functions $\chi_{HG}(\theta)$ (b) for group of particles in the angular range 0° 40° (1 - of the 1st layer, 2 - of the 2nd layer).

which is characterized by the asymmetry parameter g:

$$g = \int_0^{\pi} \sigma(\theta) sin(\theta) cos(\theta) d\theta, \tag{3}$$

where θ is the scattering angle. Further we use these results for MC numerical simulations in order to find out, whether the interference effects in rigorous $\chi(\theta)$ have influence on the radiation transfer in two-layered medium.

3.2. Monte Carlo simulation

The MC approach for numerical calculation of the intensity radiation scattered within a randomly inhomogeneous turbid medium is based on simulation of the photon trajectories N within the medium between the points of the isotropic source and the detector. MC simulation of light scattering was designed in order to solve RTE for scattering media with parameters from the table 2.

The algorithm consists of three steps [10]:

(i) Double-local estimation computation [11] (allows to obtain the scattered radiance in the single point).

(ii) Calculation of the collision point.

(iii) Calculation of scattering parameters.

3.3. Single scattering

Double local estimation in Monte Carlo numerical scheme is used to determine higher order scattered radiation. If it needs to calculate the first-order scattered radiance, we use the following equation [12]:

$$L^1(\theta) = \frac{\Lambda\varepsilon \exp(-\varepsilon r)}{4\pi r sin(\theta)} \int_0^{\pi} \chi(\eta) exp(-\varepsilon z sin(\theta)(tan(\eta/2) - \tan(\theta/2))) d\eta. \tag{4}$$

Table 2. Parameters of the scattering medium.

Layer	1	2
Single scattering albedo, Λ	1.00	1.00
Extinction coefficient, ε_{ext}	20.9	3.35
Mean free path, l	0.0478	0.2985
Layer thickness, z/l	1.00	1.00
Asymmetry parameter, g	0.5654	0.2978
Number of trajectories	5000	5000

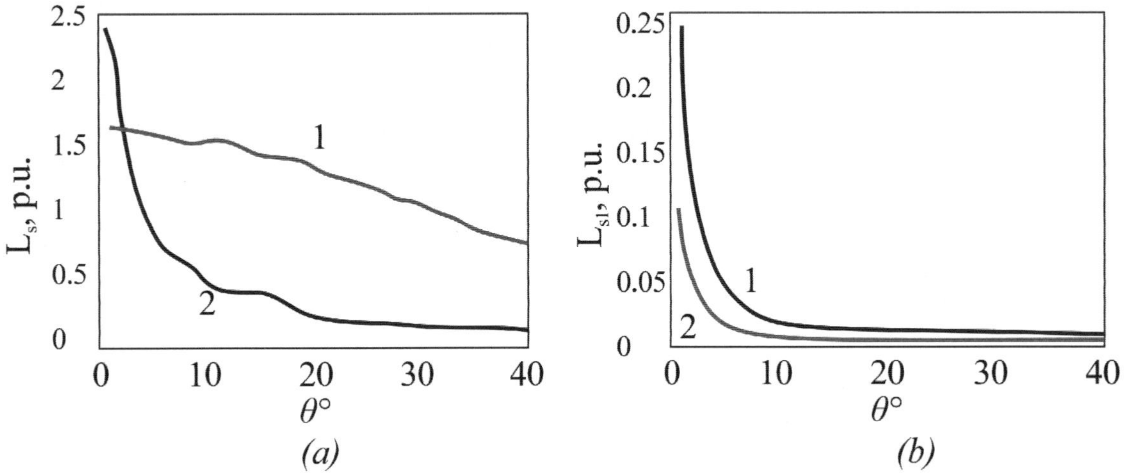

Figure 3. Multiple scattered component of angular dependence of radiance for stratified media (a), first-ordered scattered component of angular dependence of radiance for stratified media (b): using rigorous $\chi_{rig}(\theta)$ (1) and $\chi_{HG}(g,\theta)$ (2) phase functions.

In case of two-layered media it is necessary to change this equation using probabilistic approach.
1. When the single collision is in the second layer, probability is to pass without collisions through the first layer:

$$P(z_1) = exp(-\frac{z_1}{lcos(\eta)}), \tag{5}$$

Value of the single scattering expression 4 decreases proportionally to the probability:

$$L_2^1(\theta) = \frac{\Lambda\varepsilon_2 \exp(-\varepsilon_2 z_2)}{4\pi z_2 sin(\theta)} \int_0^{\pi} P(z_1)\chi_2(\eta)exp(-\varepsilon_2 z_2 sin(\theta)(tan(\eta/2) - \tan(\theta/2)))d\eta. \tag{6}$$

2. When the single collision is in the first layer, probability is to pass without collisions through the second layer:

$$P(z_2) = exp(-\frac{z_2}{lcos(\eta)}), \tag{7}$$

The value of the single scattering expression 4 decreases proportionally to the probability:

$$L_1^1(\theta) = \frac{\Lambda\varepsilon_1 \exp(-\varepsilon_1 z_1)}{4\pi z_1 sin(\theta)} \int_0^{\pi} P(z_2)\chi_1(\eta) exp(-\varepsilon_2 z_1 sin(\theta)(tan(\eta/2) - \tan(\theta/2)))d\eta. \quad (8)$$

As a result, single-scattered radiance is determined as:

$$L^1(\theta) = L_1^1(\theta) + L_2^1(\theta) \quad (9)$$

4. Conclusion

In the present work, the analysis technique of the first-order scattered radiation brightness for medium with 2 layers was shown. Comparative analysis of multiple scattered radiation brightness for medium with two layers shows the following:

- interference of the scattered waves inside the group of densely packed particles in stratified medium impacts on the overall radiance distribution, even if the layers have different parameters.
- application of model phase function with single parameter for radiation transfer calculation in stratified media leads to significant errors, which can be prevented by using the rigorously obtained phase function.

Acknowledgements

The authors are thankful for the financial support of the Russian Scientific Foundation (Project # 14-19-01083).

References

[1] Mishchenko M I and Travis L D 1997 *Journal of Geophysical Research: Atmospheres* **102** 13543–13553
[2] Yang P, Feng Q, Hong G, Kattawar G W, Wiscomb W J, Mishchenko M I, Dubovik O, Laszlo I and Sokolik I N 1995 *J. of Aerosol Sci.* **38** 995–1014
[3] Schecklman S, Zurk L M, Henry S and Kniffin G P 2011 *Journal of Applied Physics* **109** 094902
[4] Zurk L M, Orlowski B, Winebrenner D P, Thorsos E I, Leahy-Hoppa M R and Hayden L M 2007 *J. Opt. Soc. Am. B* **24** 2238–2243
[5] Kirillin M, Meglinski I, Kuzmin V, Sergeeva E and Myllylä R 2010 *Opt. Express* **18** 21714–21724
[6] Mishchenko M I and Travis L D 2006 *Multiple scattering of light by particles: radiative transfer and coherent backscattering* (New York: Cambridge Univ. Press)
[7] Ermakov S M and Mikhailov G A 1976 *Statistical modeling* (Moscow: Nauka)
[8] Henyey L C and Greenstein J L 1941 *Astrophysical Journal* **93**
[9] Taflove A and Hagness S C 2000 *Computational Electrodynamics: The Finite-Difference Time-Domain Method* (Boston: Artech House)
[10] Dolganova I N, Neganova A S, Karasik V E, Zaytsev K I and Yurchenko S O 2015 *Journal of Physics: Conference Series* **584** 012019
[11] Marchuk G I 1976 *Monte-Carlo method in atmosphere optics* (Moscow: Nauka)
[12] Savenkov V I and Mel'nikov G N 1979 *Light felds in ocean* (Moscow: AS USSR)

On a charged particle's spin evolution induced by a strong laser

A Khvedelidze[1,2,3], D Mladenov[4] and I Rogojin[3]

[1] A Razmadze Mathematical Institute, Tbilisi State University, Georgia
[2] Institute of Quantum Physics and Engineering Technologies, Georgian Technical University, Tbilisi, Georgia
[3] Laboratory of Information Technologies, Joint Institute for Nuclear Research, Dubna, Russia
[4] Faculty of Physics, Sofia State University, Sofia, Bulgaria

E-mail: akhved@jinr.ru

Abstract. The quantum and classical dynamics of a charged, spin-1/2 particle interacting with a strong laser, modeled by the elliptically polarized monochromatic electromagnetic plane wave, is considered in the semi-classical approach. The charge interaction with a high intensity radiation is described classically without using of the dipole approximation. The particle's spin evolution is treated according to the Pauli equation with relativistic corrections caused by the laser intensity. The resonance character of transitions between spin-1/2 states for certain regimes of intensities and values of polarization is described.

1. Introduction

The recent successes in a laser technologies envisage building of the coherent light sources that can provide electromagnetic field intensities in excess of $10^{20}\,\mathrm{W\,cm^{-2}}$ (cf. [1]). This opens a new perspectives in our studies of light-matter interactions, that are important for the fundamental physics as well as for diverse applications. As studies show (see e.g. [2] and references therein), the classical/quantum dynamics of a charged particle in a strong electromagnetic background turns out to be far from a customary view. The later are mainly based on the so-called electric dipole approximation [3]. However, when the radiation intensity is sufficiently large, several factors, such as the retardation effect, the role of the magnetic part of Heaviside-Lorentz force in distortion of a particle's classical orbit as well as the Thomas correction, cannot be ignored any more. The exit beyond the dipole approximation gives not only quantitative corrections to results, obtained for a small light intensity, but sometimes may lead to their qualitative changes. The present report aims to describe such a non-dipole effects in the the spin evolution of charged particle caused by its interaction with a high intensity laser.

Quantity characterizing the intensity of a coherent radiation interaction is given by the dimensionless laser field strength parameter:

$$\eta^2 := -2\,\frac{e^2}{m^2c^4}\,\langle\langle\, A_\mu A^\mu \,\rangle\rangle\,, \tag{1}$$

where m and e are particle mass and electric charge, $\langle\langle \cdots \rangle\rangle$ stands for time average and A_μ is a four vector describing the laser field [4, 5]. The quantity η^2 specifies the relativistic

Journal of Physics: Conference Series **672** (2016) 012002 doi:10.1088/1742-6596/672/1/012002

properties of a particle's dynamics. For small $\eta^2 \ll 1$, the dipole approximation works and it is well established that a charged particle's spin behaves adiabatically, linearly responding to the magnetic component of the electromagnetic field. When η^2 is increasing the dipole approximation breaks down, non-linear effects come in force and the adiabatic picture of the spin precession turns to be inadequate. Below, we briefly state a method allowing to describe the evolution of a spin-1/2 particle in a strong monochromatic plane wave for the "semi-relativistic interaction regime", when effects of η^2- order are not negligible any more, but still $\eta^2 < 1$ and expansions over η^2 are correct.

The paper is organized as follows. After formulation of a laser-particle interaction model a short exposition of its classical dynamics beyond the dipole approximation is given following [6]. Then, in accordance with the semi-classical approach, the particle's spin dynamics is investigated solving the Pauli equation, with effective magnetic field evaluated along the charge classical trajectory. Our studies reveal several non-dipole effects induced by a strong laser, each with non-trivial dependence on a laser's intensity and polarization. Below two of them, the deviation of spin precession frequency from the laser frequency and the resonance character of a charged particle's spin-flip transitions will be discussed.

2. The model

The laser radiation is modeled by a monochromatic elliptically polarized plane wave propagated along the z-axis. The corresponding gauge potential can be fixed as follows

$$\mathbf{A} := a\Big(\varepsilon\cos(\omega_L\xi)\,,\ \sqrt{1-\varepsilon^2}\sin(\omega_L\xi)\,,\ 0\Big)\,, \tag{2}$$

where $a = \mathcal{E}_0/\omega_L$, the phase $\xi := t - z/c$ and $\varepsilon \in [0,1]$ is the light polarization parameter.

We suppose that a charge carrier particle is in pure quantum state Ψ, which is decomposable in charge & spin components:

$$|\Psi\rangle = \sum_{i=0,1}\sum_{\alpha=\pm} c_{\alpha,i}|\psi_\alpha\rangle \otimes |\chi_i\rangle\,. \tag{3}$$

Two states $|\psi_\pm\rangle$ in (3) are linearly independent semi-classical solutions to the Schrödinger equation for a charge interacting with a laser radiation

$$\imath\hbar\,\frac{\mathrm{d}}{\mathrm{d}t}\,|\psi\rangle = -\frac{\hbar^2}{2m}\Big(\boldsymbol{\nabla} - \frac{\imath e}{\hbar c}\,\mathbf{A}(t,\boldsymbol{x})\Big)^2\,|\psi\rangle\,. \tag{4}$$

The semi-classical calculations imply that the state vectors $|\chi_i\rangle$, describing spin degrees in (3), satisfy the spin evolution equation:

$$\imath\hbar\,\frac{\mathrm{d}}{\mathrm{d}t}\,|\chi\rangle = \mathrm{H}(t)|\chi\rangle\,, \tag{5}$$

where

$$\mathrm{H}(t) = -\left(\frac{ge}{2mc}\left(\mathbf{B} - \frac{1}{c}\,[\boldsymbol{v}\times\mathbf{E}]\right) + \frac{1}{2c^2}\,[\boldsymbol{v}\times\dot{\boldsymbol{v}}]\right)\mathbf{S}\,. \tag{6}$$

The $\mathbf{E}$ and $\mathbf{B}$ in (6) are electric and magnetic components of a laser field evaluated along a particle's classical orbit seen in the LAB frame. The term in parenthesis is the magnetic field in the instantaneous rest frame of a charged particle, (Galilei boosted), while the last contribution in (6) is the leading relativistic Thomas precession correction [7] due to the non-vanishing curvature of a particle's trajectory. Below the spin-1/2 particle will be considered, whose spin operator is $\mathbf{S} = \hbar/2\,\boldsymbol{\sigma}$, with the standard Pauli matrices $\boldsymbol{\sigma}$.

3. Sketch of calculations

The forthcoming computations rest on the classical non-relativistic Hamilton-Jacobi problem for spinless particle moving in the electromagnetic background (2). Due to the knowledge of the exact solution to this problem [6], one can determine both, the leading semi-classical solution to the Schrödinger equation (4) as well as calculate the effective spin-laser interaction (6).

3.1. The classical dynamics

To make formulas compact let us impose the initial condition on classical trajectory $\boldsymbol{x}(0) = 0$, and fix the frame, where the average of a particle's velocity component orthogonal to the electromagnetic wave propagation direction is zero, $\langle\langle \boldsymbol{v}_\perp \rangle\rangle = 0$. According to [6] the Hamilton-Jacobi generating function for a spinless particle traveling in an arbitrary plane wave background of the form $(\boldsymbol{A}_\perp(\xi), 0)$ reads

$$\mathcal{F}(\xi\,, \boldsymbol{\Pi}) \;=\; -c(mc - \Pi_z)\,\xi + c\int_0^\xi du\,\sqrt{(mc-\Pi_z)^2 + W(u, \boldsymbol{\Pi}_\perp)}\,, \tag{7}$$

where

$$W(\xi, \boldsymbol{\Pi}_\perp) := -\frac{e^2}{c^2}\boldsymbol{A}_\perp^2 + 2\frac{e}{c}\boldsymbol{A}_\perp \cdot \boldsymbol{\Pi}_\perp\,. \tag{8}$$

Here the constants Π_z and $\boldsymbol{\Pi}_\perp$ are determined from the initial value of the particle velocity. With the aid of (7) the standard calculations give the leading semiclassical wave function [1] :

$$\langle \boldsymbol{x}, t|\psi_+\rangle = \frac{1}{\sqrt{|\partial\mathcal{F}/\partial\xi|}}\exp\frac{i}{\hbar}\left(\mathcal{E}t + \boldsymbol{\Pi}_\perp\cdot \mathbf{x}_\perp\right)\exp\frac{i}{\hbar}\mathcal{F}. \tag{9}$$

Since the generating functions (7) determines a particle's trajectory as an explicit function of laboratory frame time t, the evaluation of the effective magnetic field (6) along the classical orbit becomes straightforward. A quick glance shows that in order to compute (6) it is enough to owe from [6] the expression for a particle's velocity

$$\boldsymbol{v} = c\left(-c\eta\varepsilon\,\mathrm{cn}\,(u,\,\mu)\,, \eta\sqrt{1-\varepsilon^2}\,\mathrm{sn}\,(u,\,\mu)\,, 1-\gamma_z\,\mathrm{dn}\,(u,\,\mu)\right) \tag{10}$$

as well as the expression for its z-coordinate:

$$z(t) = ct - \frac{c}{\omega_L}\,\mathrm{am}(u,\,\mu)\,. \tag{11}$$

In (10)-(11) $\mathrm{sn}(u,\,\mu)$, $\mathrm{cn}(u,\,\mu)$, $\mathrm{dn}(u,\,\mu)$ denote the double periodic elliptic Jacobian functions, and $\mathrm{am}(u,\,\mu)$ is the Jacobian amplitude function [8]. The argument, $u := \omega'_L t$, of these functions is the laboratory frame time t scaled by the non-relativistically Doppler shifted laser frequency $\omega'_L = \gamma_z\,\omega_L\,,\quad c\gamma_z = c - v_z(0)$. The modulus μ is determined by a laser's and a particle's characteristics:[2] $\gamma_z^2\,\mu^2 = (1-2\,\varepsilon^2)\,\eta^2$. Note that, the quarter period K of the Jacobian functions is non-linear function of the modulus μ. Therefore, the fundamental circular frequency of the particle's motion, $\omega_P = 2\pi/T_P$, differs from the frequency ω_L of a laser field. According to (10) the components of a charged particle velocity in the plane orthogonal to the wave propagation are periodic functions of time with the period $T_P := 4\mathrm{K}/\omega'_L$, while in the direction of propagation the period of oscillations is twice smaller.

[1] To simplify expressions we assume that the initial state has only one non vanishing coefficient, $c_{+,0}$. The unit normalization condition for (9) fixes it $\pi c_{+,0}^2 = 2m\,\omega_P$.

[2] In (10)-(11) the modulus μ belongs to the fundamental domain, $0 < \mu^2 < 1$. The solutions outside this interval can be reconstructed using the modular properties of the Jacobian functions [6].

3.2. The spin-1/2 dynamics

Using the above described classical solutions for the charge motion, the laser-spin interaction (6) reduces to the $\mathrm{H}(t) := -\boldsymbol{\Omega}(t)\cdot\mathbf{S}$, with the following Larmor vector $\boldsymbol{\Omega}(t)$

$$\Omega_1(t) = \frac{1}{2}\eta\omega'_L\sqrt{1-\varepsilon^2}[(g+1)\mathrm{dn}(\omega'_L t\,\mu) - \gamma_z]\mathrm{cn}(\omega'_L t,\mu), \tag{12}$$

$$\Omega_2(t) = \frac{1}{2}\omega'_L\varepsilon\left[(g+1)\mathrm{dn}(\omega'_L t,\mu) - \gamma_z(1-\mu^2)\right]\mathrm{sn}(\omega'_L t\,\mu)\,, \tag{13}$$

$$\Omega_3(t) = -\frac{1}{2}\eta^2\omega_L\varepsilon\sqrt{1-\varepsilon^2}\left[g-\gamma_z\,\mathrm{dn}(\omega'_L t\,,\mu)\right]. \tag{14}$$

So far, a particle's spin-1/2 effectively interacts with a laser via the spatially homogeneous, time depending magnetic field given by a superposition of the alternating magnetic field (12)-(13) in a plane orthogonal to the wave propagation and a small, "almost constant" magnitude magnetic field (14) along the laser beam. Note that, in the leading order expansion over the powers of modulus μ expressions (12)-(14) reduce to the non-relativistic Larmor vector corresponding to the laser magnetic field.

4. Non-dipole effects examples

In this section, based on the equation (5), we make an outline of the derivation of two non-dipole effects in spin evolution due to the interaction particle interaction with a strong elliptically polarized laser: non-linear dependence of spin-precession frequency on the laser frequency, and the resonance character of spin-flip transitions.

4.1. Spin effects under the cyclic evolution

Consider at first a spin evolution in the linearly polarized laser field. In this case according to expressions (12)-(14) the spin-1/2 wave function displays non-trivial periodicity:

$$|\chi(t+4\mathrm{K}(\mu)/\omega')\rangle = |\chi(t)\rangle\,, \tag{15}$$

where $\mathrm{K}(\mu)$ is the quarter period of the Jacobian function. Apart from the pure kinematic non-relativistic Doppler shift, the particle's spin precession frequency depends non-linearly on the laser intensity and polarization through the period $\mathbb{K}$. For the semi-relativistic intensities $\eta \ll 1$, the period of the particle oscillation can be represented in the form of the expansion

$$T_P = \frac{2\pi}{\omega'}\left[1+\left(\frac{1}{2}\right)^2 1-2\,\varepsilon^2\gamma_z^2\,\eta^2 + \left(\frac{1\cdot 3}{2\cdot 4}\right)^2\frac{(1-2\,\varepsilon^2)^2}{(1-\beta_z)^4}\,\eta^4 + \ldots\right].$$

The presence of $\mathbb{K}$ in (15) exposes a fundamental peculiarity of the particle dynamics which is beyond the dipole approximation.

4.2. The spin flip resonance in a laser field

Our computations show, that in a strong laser instead of the well-known dipole approximation result, vanishing spin-flip amplitude, the resonance transition occurs for certain values of the laser field strength, η_*^2. The position of the resonance point η_*^2 depends on particle's gyromagnetic ratio and the laser field polarization.

Journal of Physics: Conference Series **672** (2016) 012002 doi:10.1088/1742-6596/672/1/012002

4.2.1. Circularly polarized laser Let us at first consider motion in the circularly polarized background. In this case an explicit form of the resonance can be derived [9]. In this case spin evolution problem reduces to the well-known Rabi problem [10] and the spin-flip probability reads

$$\mathcal{P}_{\downarrow\uparrow} = \frac{\kappa^2\eta^2}{\kappa^2\eta^2 + (\eta^2 - \eta_*^2)^2}\,\sin^2(\omega_S t)\,. \tag{16}$$

An intensity resonance occurs at $\eta_*^2 := 4/(g-1)$ and the spin probability oscillates with the frequency ω_S

$$\omega_S := \frac{\omega_L|1-g|}{8}\,\sqrt{\kappa^2\eta^2 + (\eta^2 - \eta_*^2)^2}\,, \tag{17}$$

determined not only by ω_L, but the laser intensity as well. In (16) the constant $\kappa^2 := 2\,g^2/(1-g)^2$ was introduced.

4.2.2. Elliptically polarized laser In order to study the generic case of elliptically polarised laser we adopt the method of reducing the system of linear equations (5) to the analysis of the Riccati equation (see e.g. [11]).

The evolution operator $U(t, t_0)$ for (5) can be represented in the factor form using the operators $S_\pm = 1/2\,(\sigma_1 \pm \imath\,\sigma_2)$ and $S_0 = 1/2\,\sigma_3$:

$$U(t,0) = \exp\left(a(t)S_+\right)\exp\left(b(t)S_0\right)\exp\left(c(t)S_-\right)\,. \tag{18}$$

Three unknown functions $a(t), b(t)$ and $c(t)$ determine a generic spin-1/2 state. Noting that the probabilities of transitions between states labelled by spin-1/2 projection to z-axis are given by formulas

$$\mathcal{P}_{\uparrow\uparrow} = \frac{1}{1+|a|^2}\,, \qquad \mathcal{P}_{\uparrow\downarrow} = \frac{|a|^2}{1+|a|^2}\,, \tag{19}$$

we will concentrate on the determination of function $a(t)$. From the spin evolution equation for (18) it follows that $a(t)$ is subject to the Riccati equation:

$$\dot{a} = \imath(\Omega_+ a^2 + \Omega_0 a - \Omega_-)\,. \tag{20}$$

with functions $\Omega_\pm = 1/2\,(\Omega_1 \pm \imath\,\Omega_2)$ and $\Omega_0 = 1/2\,\Omega_3$.

Our numerical studies of (20) allows to determine transition probabilities (19) for different values of radiation intensities, polarization and the particle's gyromagnetic ratio. From the curve depicted on Figure 1 one can see the dependence of spin-flip transition from a laser intensity for for fixed value of gyromagnetic ratio (g=6) and light polarization. Analysis shows that the value of intensity resonance point η_*^2 non-trivially depends on the values a laser polarization. As an example, the Figure 2 shows the *"resonance intensity curves"*, i.e., points on the $(\eta, g)-$plane for which the spin-flip occurs with the unit probability for two polarization, the circular one and elliptical with $\varepsilon = 1/2$.

5. Conclusion

In conclusion, our studies point on few spin effects associated with a spin-1/2 charged particle motion in laser fields whose intensities provide the semi-relativistic interaction regime. These effects are identified beyond the dipole approximation and could be attributed to the fact that the magnetic forces have been taken into account and alter the spin evolution appreciably.

Finally, it is worth to point out some flaws of our study. First of all, the results stated above are strongly based on the semiclassical approach used. Therefore it is necessary to convince

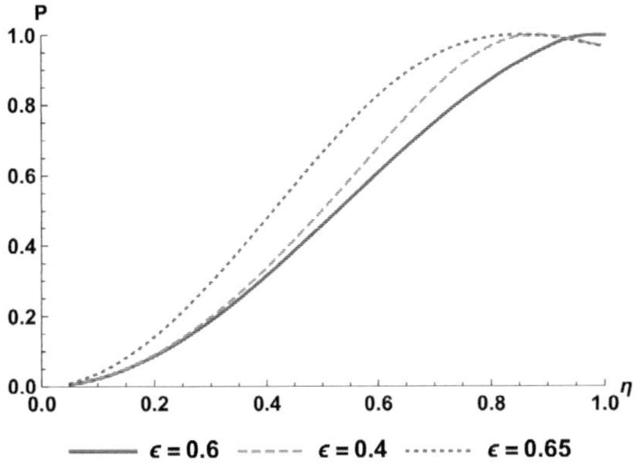

Figure 1. Spin-flip probability $\mathcal{P}$ vs. intensity η, gyromagnetic ratio g=6.

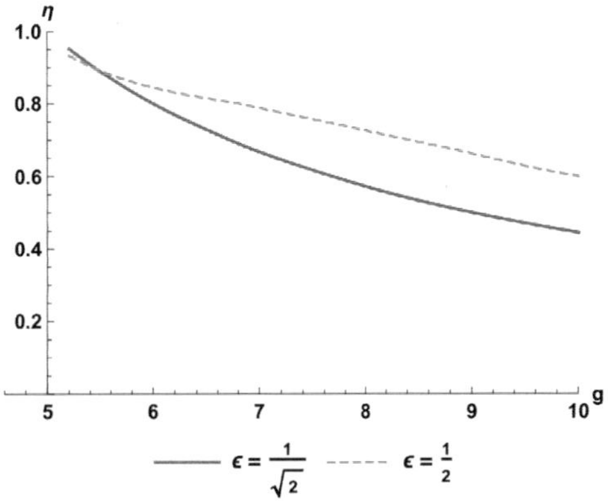

Figure 2. Typical resonance curves on (η, g)-plane for different polarisation.

that the resonant character of spin flip is not a spurious feature due to a fault approximation. A complete relativistic consideration within the Dirac theory or using the Bargmann, Michel & Telegdi spin evolution equations [12] is required. Besides this, it should be noted that the effects of a back reaction of a radiation as well spin forces distorting the spinning particle's trajectory (cf. e.g. the discussion in [13],[14]), [15] may alert our conclusions.

Acknowledgments
This work has been supported in part by the JINR-Bulgaria collaborative grant and the Ministry of Education and Science of the Russian Federation under the grant No. 3802.2012.2.

References
[1] Mourou G and Tajima T 2011 *Science* **331** 41
[2] Piazza A Di, Mller C, Hatsagortsyan K Z and Keitel C H 2012 *Rev. Mod. Phys.* **84** 1177
[3] Jackson J D 1999 *"Classical electrodynamics"* (New York: Wiley)
[4] Sengupta N 1949 *Bull. of Calcutta Math. Soc.* **41** 187
[5] Sarachik E S and Schappert G T 1970 *Phys. Rev.* **D** 1 2738
[6] Jameson P and Khvedelidze A 2008 *Phys. Rev.* **A77** 053403
[7] Thomas L H 1926 *Nature* 117 514
[8] Whittacker E T and Watson G N 1952 *A Course of Modern Analysis* (Cambridge University Press)
[9] Khvedelidze A 2008 *Proc. Int. XIII Conf. "Selected Problems of Modern Physics"*, ed B Barbashov and S Eliseev (Dubna, Russia) 365
[10] Rabi I I 1937 *Phys. Rev.* **51** 652
[11] Popov V S 2007 *Usp. Fiz. Nauk* **177** 1319
[12] Bargmann V, Michel L and Telegdi V 1959 *Phys. Rev. Lett.* **2** 435
[13] Walser M W and Keitel C H 2000 *J. Phys. B: At. Mol. Opt. Phys.* **33** L221
[14] Roman J S, Roso L and Plaja L 2004 *J. Phys. B: At. Mol. Opt. Phys.* **37** 435
[15] Wen M, Bauke H and Keitel C H 2014 Dynamical spin effects in ultra-relativistic laser pulses, *Preprint* 1406.3659

Journal of Physics: Conference Series **672** (2016) 012003 doi:10.1088/1742-6596/672/1/012003

Temperature effects on fidelity of reflection from absorbing Bragg mirrors

D Čevizović[1] and A V Chizhov[2,3]

[1]Vinča Institute of Nuclear Sciences, Belgrade, Serbia
[2]Bogoliubov Laboratory of Theoretical Physics, Joint Institute for Nuclear Research, 141980 Dubna, Russia
[3]Dubna State University, 141980 Dubna, Russia

E-mail: chizhov@theor.jinr.ru

Abstract. In the paper, we consider the quantum effects in reflection of the electromagnetic field from an absorbing Bragg mirror. The special emphasis is placed on the study of the influence of thermal noises of the mirror environment on the fidelity of reflection. Taking the incident light field in coherent and Fock states, we show that the fidelity of reflection turns out to be strongly dependent on the quantum state, and on the level of thermal noise and absorption of the mirror.

1. Introduction

A Bragg mirror (or distributed Bragg reflector) is an optical device which is composed of an alternating sequence of multiple layers of two different materials. Each optical layer thickness corresponds to one quarter of the wavelength that provides constructive interference of the reflected waves from the mirror layers. There thus appears a certain range of wavelengths (called the photonic stop band), within which light propagation in the mirror structure is strongly suppressed. It is why Bragg mirrors act as high-quality reflectors, so that they have a wide use in many optical devices, such as light-emitting diodes [1], vertical cavity surface emitting lasers [2], solar cells [3], and Fabry-Pérot filters [4].

However, in the case of the quantized electromagnetic fields the mirror reflectivity, which is just related to intensities of the incident and reflected fields, cannot fully characterize the process of reflection. This is due to the fact that the interaction of the incident electromagnetic field with the mirror materials might cause a significant change in the quantum state of reflected light. Instead, the fidelity of the incident and reflected states of light that shows their quantum closeness may serve as the measure of the reflection efficiency.

The present report is aimed to investigate the effects of reflection of quantum states from a Bragg mirror in view of thermal noises of its environment. Such an analysis looks to be important for the quantum information processing of light states with the use of optical cavities [5].

2. The Model

From the theoretical point of view a Bragg reflector can be considered as a four-port device formed as a sequence of different semiconductor and dielectric plates of the certain refractive indices (figure 1). Our description of the quantum-state transformation of the fields by the Bragg reflector is based on the input-output relations at dispersive and absorbing four-port

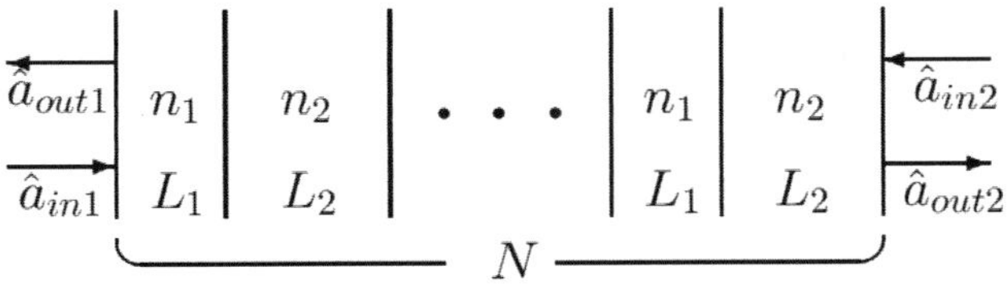

Figure 1. Four-port scheme for a N-stack Bragg mirror.

devices [6, 7]. These relations were derived by using the Green-function approach for quantization of the electromagnetic field in dispersive and absorbing media and have the form [6]

$$\begin{pmatrix} \hat{a}_{out1}(\omega) \\ \hat{a}_{out2}(\omega) \end{pmatrix} = \mathbf{T}(\omega) \begin{pmatrix} \hat{a}_{in1}(\omega) \\ \hat{a}_{in2}(\omega) \end{pmatrix} + \mathbf{A}(\omega) \begin{pmatrix} \hat{g}_{+}(\omega) \\ \hat{g}_{-}(\omega) \end{pmatrix}. \tag{1}$$

In equation (1) $\hat{a}_{in1,2}$ and $\hat{a}_{out1,2}$ stand for annihilation operators of the incoming and outgoing fields, respectively. The bosonic annihilation operators $\hat{g}_{\pm}$ describe excitations associated with the device and play the role of the additional operator noise sources. The 2×2 matrices $\mathbf{T}(\omega)$ and $\mathbf{A}(\omega)$ are the characteristic transformation and absorption matrices of the device. When the device is surrounded by vacuum, then the the matrices $\mathbf{T}(\omega)$ and $\mathbf{A}(\omega)$ obey the relation

$$\mathbf{T}(\omega)\mathbf{T}^{\dagger}(\omega) + \mathbf{A}(\omega)\mathbf{A}^{\dagger}(\omega) = \mathbf{I}, \tag{2}$$

where $\mathbf{I}$ is a unity 2×2 matrix. In such notations, $\mathcal{R} = |T_{11}(\omega)|^2$ and $\mathcal{T} = |T_{12}(\omega)|^2$ mean the reflection and transmission coefficients, respectively, whereas $\mathcal{A} = (|A_{11}(\omega)|^2 + |A_{12}(\omega)|^2) = (1 - \mathcal{R} - \mathcal{T})$ represents the coefficient of absorption.

The elements of the characteristic transformation and absorption matrices of arbitrary four-port device can be calculated in principle from its Green function. However, in case the device has a layered structure of the certain refractive indices, then these elements might be obtained analytically step by step without explicitly calculating the multislab Green function [6]. The latter method seems to be particular convenient for the Bragg reflectors which have periodic structures.

3. Quantum-state transformation

The operator input-output relations (1) determine the transformation that relates the state of the outgoing field to the state of the incoming field [7]. For the further convenience, we will describe the state transformation in the phase-space representation in terms of characteristic functions. In case the signal field is sent to the first port, the normally-ordered input and output characteristic functions then read as

$$\chi^{\mathcal{N}}_{in1}(\beta) = \mathrm{Tr}\{\hat{\rho}\exp(\beta\hat{a}^{\dagger}_{in1})\exp(-\beta^{*}\hat{a}_{in1})\}, \quad \chi^{\mathcal{N}}_{out1}(\beta) = \mathrm{Tr}\{\hat{\rho}\exp(\beta\hat{a}^{\dagger}_{out1})\exp(-\beta^{*}\hat{a}_{out1})\}, \tag{3}$$

in which $\hat{\rho}$ is the density matrix of the whole system and β is a complex variable designating a point in the phase space. Taking into account the operator relation (1), we arrive at the following form for the output characteristic function

$$\chi^{\mathcal{N}}_{out1}(\beta) = \chi^{\mathcal{N}}_{in1}(T^{*}_{11}\beta)\chi^{\mathcal{N}}_{in2}(T^{*}_{12}\beta)\chi^{\mathcal{N}}_{g_{+}}(A^{*}_{11}\beta)\chi^{\mathcal{N}}_{g_{-}}(A^{*}_{12}\beta) \tag{4}$$

given in terms of normally-ordered characteristic functions of the input radiation fields $\chi^{\mathcal{N}}_{in1,2}$ and of device excitations $\chi^{\mathcal{N}}_{g\pm}$ scaled by the corresponding elements of the characteristic transformation and absorption matrices.

In order to determine the quantum closeness of incident and reflected radiation fields, we use the fidelity which may serve as the measure of the quantum reflection efficiency. If the incident field is in a pure quantum state, in terms of the normally-ordered characteristic functions the fidelity can be represented as [8]

$$\mathcal{F} = \frac{1}{\pi} \int d^2\beta \chi_{in1}^{\mathcal{N}}(-\beta) \chi_{out1}^{\mathcal{N}}(\beta) \exp(-|\beta|^2). \tag{5}$$

4. Temperature effects

To find the influence of the mirror environment on the quantum reflection efficiency, we ought to specify the quantum states of the mirror excitations. As soon as we are interested in the temperature effects, we have to assume that the mirror excitations are in chaotic states at certain temperature T. It means that their normally-ordered characteristic functions are

$$\chi_{g_\pm}^{\mathcal{N}}(\beta) = \exp(-\overline{n}_{th}|\beta|^2), \tag{6}$$

where $\overline{n}_{th} = [\exp(\hbar\omega/k_B T) - 1]^{-1}$ denotes the mean number of the mirror excitations. On purpose to eliminate an active interference of the second port in the reflection process from the first one, we consider the radiation field $\hat{a}_{in2}$ to be in the vacuum state, so that its normally-ordered characteristic function is simply to be

$$\chi_{a_{in2}}^{\mathcal{N}}(\beta) = 1. \tag{7}$$

Then the fidelity (5) takes the form

$$\mathcal{F}_{th} = \frac{1}{\pi} \int d^2\beta \chi_{in1}^{\mathcal{N}}(-\beta) \chi_{in1}^{\mathcal{N}}(\sqrt{\mathcal{R}}\beta) \exp\{-(1 + \overline{n}_{th}\mathcal{A})|\beta|^2\} \tag{8}$$

that reveals the temperature dependence only in the case of an absorbing mirror with the certain coefficient of absorption $\mathcal{A}$.

5. Results

As an example, we consider a Bragg mirror consisting of repeated pairs of titanium dioxide and silica. The number of the stacks we set to be $N = 12$. For the qualitative analysis, we neglect the dependence of the refractive indices of temperature and fix them as $n_1 = 2.5 + i3 \cdot 10^{-2}$ for TiO_2 and $n_2 = 1.5$ for SiO_2. The widths of the pair slabs L_1 and L_2 in the stack are chosen to fulfill the Bragg condition for the center wavelength of the stop band as $L_1/L_2 = \mathrm{Re}(n_2/n_1) = 1.5/2.5$. The behavior of the coefficients of reflection and absorption for such a mirror as functions of the scaled wave number is given in figure 2. The shape of the curve for the reflection coefficient is seen to demonstrate the presence of a stop band.

We then study two cases of quantum states of the signal field which are exemplary for optical processes, namely, coherent and Fock states.

The choice of the incident field in the coherent state $|\alpha_0\rangle$, which can be realized by laser, sets its normally-ordered characteristic function to be

$$\chi_{a_{in1}}^{\mathcal{N}(coh)}(\beta) = \exp(\beta\alpha_0^* - \beta^*\alpha_0). \tag{9}$$

On substituting the expression (9) into equation (8), we obtain that the fidelity of the coherent field reflection from the mirror at the temperature T is represented in the form

$$\mathcal{F}_{coh} = \frac{1}{1 + \overline{n}_{th}\mathcal{A}} \exp\left\{-\frac{(1 - \sqrt{\mathcal{R}})^2}{1 + \overline{n}_{th}\mathcal{A}}|\alpha_0|^2\right\}. \tag{10}$$

Equation (10) shows that thermal noises decrease the fidelity of the coherent field reflection from an absorbing mirror. Moreover, this decrease becomes more pronounced for the intense laser

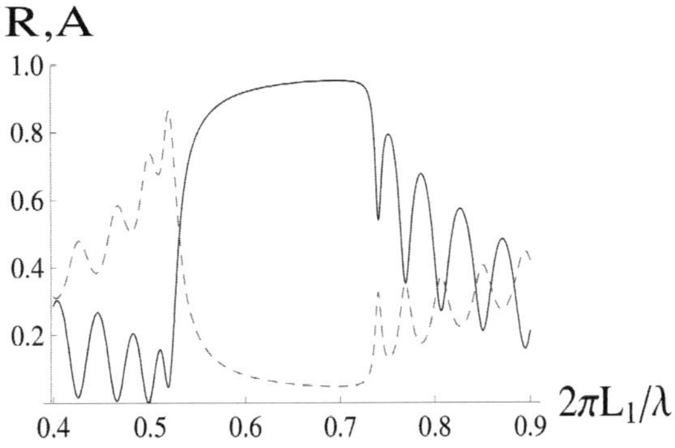

Figure 2. Coefficients of reflection $\mathcal{R}$ (solid) and absorption $\mathcal{A}$ (dashed) as functions of the radiation wave number $2\pi/\lambda$ scaled by the width L_1.

beam, i.e. when $|\alpha_0|^2$ is rather large. It leads to an effective reduction of the mirror stop band. Figure 3 illustrates such a behavior of the fidelity of the coherent field reflection on the model parameters.

If the incident field is in the n-photon Fock state, its normally-ordered characteristic function takes the form

$$\chi_{a_{in1}}^{\mathcal{N}(Fock)}(\beta) = L_n(|\beta|^2). \tag{11}$$

where L_n is the Laguerre polynomial of the n-th order. Using equation (8), we find that the fidelity of reflection of the n-photon Fock state from the mirror at the temperature T reads as

$$\mathcal{F}_{Fock} = \frac{(\overline{n}_{th}\mathcal{A} - \mathcal{R})^n}{(1 + \overline{n}_{th}\mathcal{A})^{n+1}} P_n \left(1 + \frac{2\mathcal{R}}{(\overline{n}_{th}\mathcal{A} - \mathcal{R})(1 + \overline{n}_{th}\mathcal{A})}\right), \tag{12}$$

where P_n stands for the Legendre polynomial of the n-th order. The fidelity (12) also decreases with an increase of the coefficient of absorption $\mathcal{A}$ and the level of thermal noise $\overline{n}_{th}$, as well as it displays a considerable degradation for large values of a photon number n (see figure 4). It is worth noting that this dependence turns out to be much stronger than that for the coherent state. All this might greatly reduce the quality of the reflected field in the Fock state.

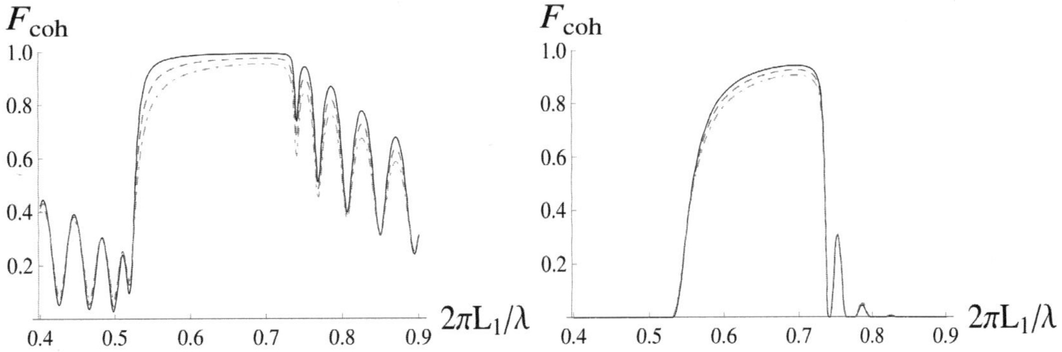

Figure 3. Fidelities of reflection for the coherent states with $\alpha_0 = 2$ (left) and $\alpha_0 = 10$ (right) when the mean number of the mirror excitations $\overline{n}_{th}$ equals 0.1 (solid), 0.5 (dashed), and 1.0 (dot-dashed).

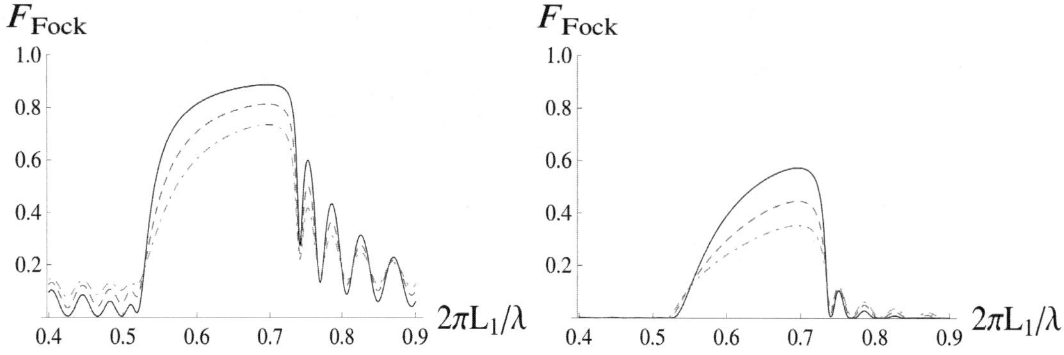

Figure 4. Fidelities of reflection for the Fock states with $n = 2$ (left) and $n = 10$ (right) when the mean number of the mirror excitations $\overline{n}_{th}$ equals 0.1 (solid), 0.5 (dashed), and 1.0 (dot-dashed).

6. Conclusion

In this work, we considered quantum effects of reflection of radiation field states from an absorbing Bragg mirror. Our description of the filed transformation by the Bragg reflector is based on the input-output relations at a four-port device obtained in the Green-function approach for quantization of the electromagnetic field in dispersive and absorbing media. As a measure of reflection, we used the fidelity of incident and reflected states of light. We investigated the dependence of fidelity of reflection with the special emphasis on thermal noises of its environment. Taking as an example coherent and Fock states of the incident light, we found that thermal noises and intensive fields of light might strongly decrease the fidelity of reflection. Moreover, for the highly nonclassical Fock states the decrease of the fidelity is appeared to be very significant. This fact ought to be taken into account when considering optical cavity QED systems.

Acknowledgments

This work was partly supported by the bilateral project between the Serbian Ministry of Education and Science and JINR (Dubna) “Theory of Condensed Matter”, and the Serbian Ministry of Education and Science under Grant No. III-45010. One of the authors (A.V.Ch.) is grateful to the Ter-Antonyan–Smorodinsky Program of the JINR (Dubna) and to the Organizing Committee of the 2nd International Symposium OPTICS-2014 for the financial support for participation in the Symposium.

References

[1] Yang Yibin *et al* 2014 *Appl. Phys. Express* **7** 042102
[2] Islam Arnob and Islam Saiful 2014 *American Journal of Optics and Photonics* **2** 37
[3] Sheng Xing, Johnson S G, Broderick L Z, Michel J and Kimerling L C 2012 *Appl. Phys. Lett.* **100** 111110
[4] Cho E, Pavlidis D and Sillero E 2007 *Phys. Stat. Sol. (C)* **4** 2764
[5] Zoller P *et al* 2005 *Eur. Phys. J. D* **36** 203
[6] Gruner T and Welsch D-G 1996 *Phys. Rev. A* **54** 1661
[7] Knöll L, Scheel S, Schmidt E, Welsch D-G and Chizhov A V 1999 *Phys. Rev. A* **59** 4716
[8] Schumacher B 1996 *Phys. Rev. A* **54** 2614

Cooperative quantum electrodynamical processes in an ellipsoidal cavity

G Alber and N Trautmann

Institut für Angewandte Physik, Technische Universität Darmstadt, D-64289, Germany

E-mail: nils.trautmann@physik.tu-darmstadt.de

Abstract. We investigate spontaneous photon emission and absorption processes of two two-level atoms trapped close to the focal points of an ellipsoidal cavity, thereby taking into account the full multimode scenario. In particular, we calculate the excitation probabilities of the atoms by describing the field modes semiclassically. Based on this approach, we express the excitation probabilities by a semiclassical photon path representation. Due to the special property of an ellipsoidal cavity of having two focal points, we are able to study interesting intermediate instances between well-known quantum-optical scenarios. Furthermore, the semiclassical photon path representation enables us to address the corresponding retardation effects and causality questions in a straightforward manner.

1. Introduction

During the last decades, the field of quantum optics has witnessed remarkable experimental developments. They have enabled new possibilities of studying resonant light-matter interaction [1–3]. These developments are not only interesting from a fundamental point of view, but also from an applied perspective, because they are relevant for advanced quantum technologies and their possible application in quantum information processing.

Understanding light-matter interaction is crucial for being able to transfer quantum information between single photons, frequently used as flying qubits, and elementary material systems, typically used as stationary qubits in applications in quantum information. Such a transfer of quantum information between single photons and elementary material systems, such as two level systems serving as qubits, is an integral building block for achieving quantum communication over large distances. Recently, considerable experimental effort [4–6] has been devoted to investigating the interaction of matter qubits with one or a few selected modes of the radiation field within the framework of the Jaynes-Cummings-Paul model [7, 8]. Recent experimental work has extended these quantum electrodynamical scenarios to the opposite limit of extreme multimode scenarios [9–11] with structured continua of electromagnetic field modes characteristic for half open cavities, such as a parabolic mirror[12–14].

Motivated by these latter developments, we explore in this paper basic dynamical features of quantum electrodynamic processes in extreme multimode scenarios. In particular, we investigate the almost resonant interaction between two two-level systems located at the focal points of a prolate-ellipsoidal cavity and the quantized electromagnetic field. A particularly interesting elementary quantum electrodynamical situation arises if these two atoms exchange a single photon emitted spontaneously by one of these two-level systems. Due to the special property of

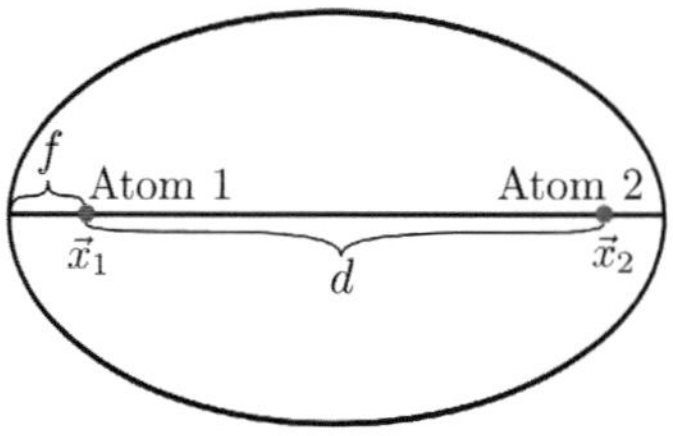

Figure 1. Two two-level systems located in the foci of a prolate-ellipsoidal cavity.

the prolate-ellipsoidal cavity of having two focal points, the electromagnetic field strength around the focal points is enhanced significantly, thus causing an interesting interaction between the two-level systems and the radiation field. From the point of view of the two-level systems as an open quantum system, the resulting reduced dynamics is highly non-Markovian with significant memory effects. Thus, an ellipsoidal cavity is an interesting scenario to study almost resonant matter field interaction. In particular, by changing the size of the cavity continuously, all intermediate cases between the single mode scenario, as described by the Jaynes-Cummings-Paul model, the large cavity limit, and the case of a structured continua of field modes can be addressed. Even the recently explored case of a parabolic cavity [15] can be reproduced in the limit of infinitely separated focal points. Furthermore, we are able to tune the coupling strength between the two-level systems and the radiation field by changing the shape of the cavity. In addition, we are able to investigate diffraction effects in the regime of wavelengths large in comparison to the characteristic length scales inside the cavity and their extinction in the geometric optical limit of small wavelengths. In the following it is demonstrated that all these effects can be described adequately with the help of a semiclassical description of the mode structure inside the ellipsoidal cavity. We show that these semiclassical methods lead to a convenient photon path representation by which all relevant quantum mechanical transition amplitudes are expressed as a linear superposition of contributions of relevant photon paths inside the cavity.

This paper is organized as follows. In Sec. 2 we introduce our theoretical model and the main approximations involved. The Helmholtz equation with the appropriate boundary conditions is solved by semiclassical methods in Sec. 3. The photon path representation for describing the time evolution of relevant quantum mechanical transition amplitudes is presented in Sec. 4. Numerical results for the dynamics of the two-level systems are discussed in Sec. 5.

2. Quantum electrodynamical model

We investigate the dynamics of two identical two-level systems, e.g. atoms or ions, situated at the focal points $\mathbf{x}_a$ ($a \in \{1, 2\}$) of an ideally conducting prolate-ellipsoidal cavity as shown schematically in figure 1. Both atoms interact with the quantized radiation field inside this cavity. The two two-level systems are assumed to be trapped in such a way that their center of mass motion is negligible. We assume that the dipole matrix elements $\mathbf{d}_a = \langle e|_a \hat{\mathbf{d}}_a |g\rangle_a$ are oriented along the symmetry axis, i.e. $z-$axis, of the system so that we may write

$$\mathbf{d}_a = D\mathbf{e}_z \text{ with } D \in \mathbb{R}. \tag{1}$$

The identical transition frequencies ω_{eg} of the two-level systems are assumed to be in the optical frequency regime and to dominate the coupling to the radiation field as measured by relevant Rabi frequencies, for example. Furthermore, the sizes of the electronic states involved in the transitions of the two-level systems are supposed to be small compared to the corresponding wavelength $\lambda_{\text{eg}} = 2\pi c_0/\omega_{eg}$ of the transition. (c_0 being the speed of light in vacuum.) Thus, the dipole and the rotating-wave approximation (RWA) are applicable. Therefore, the quantum electrodynamical interaction between the two two-level systems and the quantized electromagnetic field inside the cavity is described by the Hamiltonian

$$\hat{H} = \hat{H}_{\text{atoms}} + \hat{H}_{\text{field}} + \hat{H}_{\text{i}} \tag{2}$$

with $\hat{H}_{\text{field}} = \sum_i \hbar\omega_i \hat{a}_i^\dagger \hat{a}_i$, $\hat{H}_{\text{atoms}} = \hbar\omega_{eg} \sum_{a=1}^{2} |e\rangle_a \langle e|_a$, $\hat{H}_{\text{i}} = -\sum_{a=1}^{2} \hat{\mathbf{E}}_\perp^+(\mathbf{x}_a)\cdot\hat{\mathbf{d}}_a^- + \text{H.c.}$, and with the dipole transition operator $\hat{\mathbf{d}}_a^- = \mathbf{d}_a^* |g\rangle_a \langle e|_a$ of the two-level atom $a \in \{1,2\}$. In the Schrödinger picture the positive frequency part of the (transversal) electromagnetic field operator is given by

$$\hat{\mathbf{E}}_\perp^+(\mathbf{x}) = -\mathrm{i}\sum_i \sqrt{\frac{\hbar\omega_i}{2\epsilon_0}}\mathbf{g}_i(\mathbf{x})\hat{a}_i^\dagger \tag{3}$$

with $\mathbf{g}_i(\mathbf{x})$ denoting the orthonormal transversal electric mode functions, i.e. $\int_V d^3\mathbf{x}\ \mathbf{g}_i^*(\mathbf{x}) \cdot \mathbf{g}_j(\mathbf{x}) = \delta_{ij}$, which are assumed to fulfill the boundary conditions for an ideally conducting cavity.

3. Mode functions of the cavity and their semiclassical approximation

3.1. Separation ansatz for the relevant mode functions

In order to determine the electric mode functions by solving the Helmholtz equation, we are going to use prolate-ellipsoidal coordinates. The connection between the prolate-ellipsoidal coordinates and the Cartesian coordinates is given by

$$\begin{pmatrix} x \\ y \\ z \end{pmatrix} = \frac{d}{2}\begin{pmatrix} \cos(\varphi)\sqrt{(1-\eta^2)(\xi^2-1)} \\ \sin(\varphi)\sqrt{(1-\eta^2)(\xi^2-1)} \\ \eta\xi \end{pmatrix}. \tag{4}$$

Thereby, the focal points have been chosen to be located at the positions $\pm\mathbf{e}_z d/2$ and d denotes their distance. The ranges of the prolate-ellipsoidal coordinates are given by $\varphi \in [0, 2\pi)$, $\eta \in [-1, 1]$, and $\xi \in [1, \infty)$. In particular, the focal points have the coordinates $\xi = 1$, $\eta = \pm 1$. In these new coordinates an ellipsoid with the focal length f is defined by the surface

$$\xi = 2f/d + 1\,. \tag{5}$$

Only those mode functions enter the interaction Hamiltonian $\hat{H}_{\text{i}}$ which have a nonvanishing scalar product with the dipole transition operators of the two two-level systems situated in the focal points of the cavity. By taking into account the transversality of the radiation field, it turns out that for dipole matrix elements oriented along the z-axis all these modes can be obtained by a separation ansatz. This separation ansatz is given by

$$\mathbf{g}_i(\varphi, \xi, \eta) = \nabla \times H,\ H = \frac{1}{\mathcal{N}}\mathbf{e}_\varphi \mathcal{F}_i(\kappa, \xi)\mathcal{G}_i(\kappa, \eta)\,, \tag{6}$$

with

$$\left[\alpha_n\kappa - \kappa^2\left(\eta^2 - 1\right) - \frac{1}{1-\eta^2}\right]\mathcal{G}(\kappa,\eta) + \frac{d}{d\eta}(1-\eta^2)\frac{d}{d\eta}\mathcal{G}(\kappa,\eta) = 0\,, \tag{7}$$

$$\left[\alpha_n\kappa - \kappa^2\left(\xi^2 - 1\right) - \frac{1}{1-\xi^2}\right]\mathcal{F}(\kappa,\xi) + \frac{d}{d\xi}(1-\xi^2)\frac{d}{d\xi}\mathcal{F}(\kappa,\xi) = 0\,, \tag{8}$$

Journal of Physics: Conference Series **672** (2016) 012004 doi:10.1088/1742-6596/672/1/012004

whereby $\kappa = \omega d/(2c_0)$, α_n being the constant of separation and $\mathcal{N}$ being the normalization factor which has to be chosen such that

$$\int_V \mathbf{g}_i(x) \cdot \mathbf{g}_j^*(x) d^3x = \delta_{i,j} \; \forall i,j \; . \tag{9}$$

The index n indicates that $\mathcal{G}(\kappa, \eta)$ remains finite for $\eta \to \pm 1$ only for countable sets of the separation constant α_n. The differential equations (7) and (8) are actually identical, and the functions $\mathcal{G}$ and $\mathcal{F}$ just differ by the domains for η and ξ. The solutions of these differential equations are prolate-spheroidal wave functions [16].

3.2. Uniform semiclassical approximation

In the previous section, we have expressed the mode functions of the electric field operator in terms of prolate-spheroidal wave functions. We can obtain an analytical solution for this equations in the special case of $\alpha_n = 0$

$$\mathcal{G}(\kappa, \eta)\Big|_{\alpha_n=0} = \frac{\sin[\kappa(1-\eta)]}{\sqrt{1-\eta^2}}, \tag{10}$$

$$\mathcal{F}(\kappa, \xi)\Big|_{\alpha_n=0} = \frac{\sin[\kappa(\xi-1)]}{\sqrt{\xi^2-1}} \; . \tag{11}$$

A general closed analytical expression for arbitrary values is not known. However, with the help of semiclassical methods, approximate expressions can be found. It turns out that the mode functions with $\alpha_n \approx 0$ yield the dominant contributions. Therefore, we can exploit the exact analytical solutions for $\alpha_n = 0$ in order to improve our approximation. It is convenient to choose the following normalization conditions for the functions $\mathcal{F}(\kappa, \xi)$ and $\mathcal{G}(\kappa, \eta)$

$$\begin{aligned} \mathcal{F}(\kappa, \xi) &\underset{\kappa\xi \to \infty}{\longrightarrow} \frac{1}{\xi} \cos\left[\kappa\xi - \frac{1}{2}(n+1)\pi\right], \\ \mathcal{G}(\kappa, 0) &= (-1)^{(n+1)/2} \text{ in case of odd } n, \\ \frac{d}{d\eta}\mathcal{G}(\kappa, \eta)\Big|_{\eta=0} &= (-1)^{n/2}\kappa \text{ in case of even } n. \end{aligned} \tag{12}$$

These conditions are fulfilled by the exact solutions in equation (10) and equation (11).

In order to circumvent problems originating from the singularities of the differential equations (7) and (8), we first of all apply a transformation which removes these singularities at the positions of the focal points. Let us demonstrate this procedure by concentrating on the functions $\mathcal{F}(\kappa, \xi)$. The procedure for the functions $\mathcal{G}(\kappa, \eta)$ is analogous because the differential equations for $\mathcal{F}$ and for $\mathcal{G}$ coincide. A transformation removing the singularities is given by

$$\mathcal{F}(\kappa, \xi) = \frac{1}{\sqrt{\xi^2-1}} f(\sqrt{\xi^2-1}) \; . \tag{13}$$

Consequently, the function $f(x)$ with $x \in [0, \infty)$ is a solution of the differential equation

$$0 = \kappa^2 x \left(x^2 - \frac{\alpha_n}{\kappa}\right) f(x) - f'(x) + x\left(1+x^2\right) f''(x) \tag{14}$$

with $x = \sqrt{\xi^2 - 1}$. The focal points correspond to $x = 0$. For the determination of the relevant mode functions, we are only interested in the regular (physical) solution of equation (14) which remains finite for all possible values of $x \geq 0$.

A uniform semiclassical JWKB-approximation [17] for the solution of equation (14) can be constructed by using the differential equation

$$\kappa^2\sigma\left(\sigma^2-\frac{\alpha_n}{\kappa}\right)\tilde{f}(\sigma)-\tilde{f}'(\sigma)+\sigma\tilde{f}''(\sigma)=0\,, \tag{15}$$

as a comparison equation with the exact regular solution

$$\tilde{f}(\sigma)=e^{-\mathrm{i}\kappa\sigma^2/2}\kappa\sigma^2\,{}_1F_1\left(1-\mathrm{i}\alpha_n/4;2;i\kappa\sigma^2\right)\,. \tag{16}$$

In equation (16) ${}_1F_1$ denotes Kummer's confluent hypergeometric function [18]. Accordingly, we are going to solve equation (14) by constructing a sufficiently smooth mapping between the variables x and σ in such a way that fulfillment of equation (14) is approximately equivalent to fulfillment of equation (15). For this purpose, we transform equation (14) into a canonical form by eliminating the terms involving first order derivates. This is achieved by the transformation $f(x)=\phi(x)u(x)$ with

$$\phi(x)=\frac{\sqrt{x}}{(1+x^2)^{1/4}}. \tag{17}$$

The resulting canonical form of equation (14) is given by

$$u''(x)+\chi(x)u(x) \;=\; 0 \tag{18}$$

with

$$\chi(x)=\frac{2x^2\left(2\kappa\left(x^2+1\right)\left(\kappa x^2-\alpha_n\right)-3\right)-3}{4\left(x^3+x\right)^2}\,. \tag{19}$$

In order to find a uniform JWKB-approximation for equation (18), we have to chose an appropriate comparison equation whose solution is known exactly. For our purposes, we chose the comparison equation

$$\tilde{u}''(\sigma)+\Pi(\sigma)\tilde{u}(\sigma) \;=\; 0 \tag{20}$$

with

$$\Pi(\sigma)=\kappa^2\sigma^2-\alpha_n\kappa-\frac{3}{4\sigma^2}. \tag{21}$$

This comparison equation is related to the differential equation (15) by the relation $\tilde{f}(\sigma)=\sqrt{\sigma}\tilde{u}(\sigma)$. Furthermore, we have to choose a smooth mapping between the independent variables x and σ determined by the relation

$$\frac{d\sigma}{dx} \;=\; \sqrt{\frac{\chi(x)}{\Pi(\sigma(x))}}\,. \tag{22}$$

The initial condition for the solution of equation (22) has to be chosen such that the first positive zero of $\chi(x)$, i.e. x_0, is mapped onto the first positive zero of $\Pi(\sigma)$, i.e. σ_0, in order to avoid singularities. Solving equation (22) analytically is a complicated problem. However, simple expressions are available for $x\to 0$ and $x\to\infty$ which are given by $\sigma(x)\underset{x\to 0}{\longrightarrow} x$ and $\sigma(x)\underset{x\to\infty}{\longrightarrow}\sqrt{2x}$. The behavior for $x\to 0$ is of importance in order to evaluate the mode function near the focal points and $x\to\infty$ is relevant for the normalization (equation (12)). Thus, the JWKB-approximation for $\mathcal{F}(\kappa,\xi)$ is finally given by

$$\mathcal{F}(\kappa,\xi)=\frac{1}{(\xi^2(-1+\xi^2))^{1/4}}\left(\frac{\Pi(\sigma(\sqrt{\xi^2-1}))}{\chi(\sqrt{\xi^2-1})}\right)^{1/4}\psi(\sigma(\sqrt{\xi^2-1})) \tag{23}$$

with [18]

$$\psi(\sigma) = e^{-\pi\alpha_n/8}\sqrt{\frac{\pi\alpha_n}{16\sigma}\text{csch}\left(\frac{\pi\alpha_n}{4}\right)}\tilde{f}(\sigma). \tag{24}$$

This expression for $\mathcal{F}(\kappa,\xi)$ fulfills the normalization condition equation (12). The normalized mode function at the first focal point is now given by

$$\begin{aligned}\mathbf{g}_i(\varphi,\xi,\eta)\Big|_{\eta=1,\xi=1} &= \\ \mathbf{e}_z\frac{4}{d\mathcal{N}(\kappa,\alpha_n)}&\left(\lim_{\xi\to 1}\frac{\mathcal{F}(\kappa,\xi)}{\sqrt{\xi^2-1}}\right)\left(\lim_{\eta\to 1}\frac{\mathcal{G}(\kappa,\eta)}{\sqrt{1-\eta^2}}\right) \\ &= \mathbf{e}_z\frac{\kappa^2\pi}{4d\mathcal{N}(\kappa,\alpha_n)}\alpha_n\text{csch}\left(\pi\alpha_n/4\right). \end{aligned} \tag{25}$$

By exploiting the symmetry of the problem, we can evaluate the mode function at the second focal point and obtain

$$\mathbf{g}_i(\varphi,\xi,\eta)\Big|_{\eta=-1,\xi=1} = (-1)^n\mathbf{g}_i(\varphi,\xi,\eta)\Big|_{\eta=1,\xi=1}. \tag{26}$$

In order to normalize the mode functions according to equation (9) we have to determine the constant $\mathcal{N}(\kappa,\alpha_n)$, i.e.

$$\mathcal{N}(\kappa,\alpha_n) = \kappa\sqrt{d\pi(I^2_{\mathcal{F}}I^0_{\mathcal{G}} - I^0_{\mathcal{F}}I^2_{\mathcal{G}})} \tag{27}$$

with $(p = 0, 2)$

$$I^p_{\mathcal{G}} = 2\int_0^1 \mathcal{G}(\kappa,\eta)^2\eta^p d\eta \text{ and } I^p_{\mathcal{F}} = \int_1^{1+2f/d} \mathcal{F}(\kappa,\xi)^2\xi^p d\xi. \tag{28}$$

According to equations (25) and (26) at the focal points the normalized mode functions are determined by the function $\mathcal{N}(\kappa,\alpha_n)$ and the function $\alpha_n\text{csch}\left(\pi\alpha_n/4\right)$. The function $\alpha_n\text{csch}\left(\pi\alpha_n/4\right)$ significantly deviates from 0 only in the region around $\alpha_n = 0$. It turns out that in this region $\mathcal{N}(\kappa,\alpha_n)$ is slowly varying in comparison with the function $\alpha_n\text{csch}\left(\pi\alpha_n/4\right)$ in case of $\lambda_{eg} \ll d, f$. This can be verified by using the semiclassical potential $\chi(x)$ and a corresponding potential for the semiclassical treatment of $\mathcal{G}$. Thus the modes with $\alpha_n \approx 0$ are the ones of main importance as far as their coupling to the dipoles of the two-level systems is concerned and $\mathcal{N}(\kappa,\alpha_n)$ can be assumed to be independent of α_n. In addition it is also well-known that the relevant modes have the property $\omega \approx \omega_{eg}$ which directly translates to $\kappa \approx \kappa_{eg} = 2\omega_{eg}d/c_0$. By incorporating these facts we can approximate $\mathcal{N}(\kappa,\alpha_n)$ by $\mathcal{N}(\kappa_{eg},0)$ in the subsequent calculation.

3.3. Semiclassical quantization functions

In the previous subsection, we have identified the modes in the region of $\alpha_n \approx 0$, $\kappa \approx \kappa_{eg}$ as the ones of main importance for the dynamics of the system. However, in the case of $\alpha_n = 0$, equations (7) and (8) can be solved analytically. This particular feature of the mode functions can be exploited for incorporating the boundary conditions of an ideally conducting metallic boundary and determining the corresponding quantization functions semiclassically. These quantization functions $n_1(\alpha_n,\kappa)$, and $n_2(\alpha_n,\kappa)$ are defined in such a way that the condition

$$(n_1(\alpha_n,\kappa), n_2(\alpha_n,\kappa)) \in \mathbb{N}_0 \times \mathbb{N}_0 \tag{29}$$

determines the possible mode functions which fulfill the boundary conditions.

The quantization function $n_1(\alpha_n, \kappa)$ is determined by imposing the condition that the function $\mathcal{G}(\kappa, \eta)$ has to remain finite for $\eta \to \pm 1$. The quantization function $n_2(\alpha_n, \kappa)$ takes into account the boundary conditions of an ideally conducting surface of the cavity. In particular, this implies that the tangential components of the electric field strength and the normal components of the magnetic field strength have to vanish at the boundary which yield the constraint

$$0 = \frac{\partial}{\partial \xi} \left[\sqrt{\xi^2 - 1} \mathcal{F}(\kappa, \xi) \right]_{\xi = 2f/d+1} . \tag{30}$$

As mentioned above we have to evaluate the quantization functions around $\alpha_n \approx 0$ and $\kappa \approx \kappa_{eg}$. Therefore, we can approximate the quantization functions $n_1(\alpha_n, \kappa)$, $n_2(\alpha_n, \kappa)$ by their first order Taylor expansions in α_n and κ around the values $\alpha_n = 0$ and $\kappa = \kappa_{eg}$. This Taylor expansion is given by

$$n_i(\alpha_n, \kappa) \approx n_i(0, \kappa_{eg}) + \partial_{\alpha_n} n_i \alpha_n + \partial_\kappa n_i \left(\kappa - \kappa_{eg} \right) \tag{31}$$

with $\partial_{\alpha_n} n_i$ and $\partial_\kappa n_i$ denoting the partial derivatives of the function $n_i(\alpha_n, \kappa)$ with respect to the variable α_n respectively κ for $\alpha_n = 0$ and $\kappa = \kappa_{eg}$. By exploiting the exact analytical expressions for $\mathcal{G}(\kappa, \eta)$ and $\mathcal{F}(\kappa, \xi)$ for $\alpha_n = 0$ we obtain the relations

$$n_1(0, \kappa_{eg}) + \partial_\kappa n_1 \left(\kappa - \kappa_{eg} \right) = \frac{2\kappa}{\pi} \tag{32}$$

and

$$n_2(0, \kappa_{eg}) + \partial_\kappa n_2 \left(\kappa - \kappa_{eg} \right) = \frac{2\kappa f}{\pi d} + \frac{1}{2} . \tag{33}$$

In order to specify the quantization functions in the linearizion approximation completely, we still have to determine $\partial_{\alpha_n} n_i$. This is achieved by invoking the semiclassical quantization condition. In case of a simple JWKB-approximation with Langer substitution [17] the quantization function and the function $\mathcal{G}(\eta)$ are given by

$$n_1 = 2S(0)/\pi - \frac{1}{2} \tag{34}$$

$$\mathcal{G}(\eta) = \sqrt[4]{\frac{V(0)}{V(\eta)}} \frac{1}{\sqrt{1+\eta}} \sin(S(\eta) + \frac{\pi}{4}) \tag{35}$$

with

$$S(\eta) = \kappa \int_{\eta}^{\eta_{\text{turn}}} \sqrt{-V(\eta)} \frac{1}{1-\eta} d\eta \tag{36}$$

denoting the semiclassical eikonal equation. The quantity

$$V(\eta) = \frac{1}{4\kappa^2} - \frac{(\eta - 1)(\eta^2 - 1 - \alpha_n/\kappa)}{1+\eta} \tag{37}$$

is the semiclassical potential and η_{turn} denotes the turning point of the semiclassical potential $V(\eta)$ in the region $[0, 1]$ ($V(\eta_{\text{turn}}) = 0$). Thereby, we obtain

$$\partial_{\alpha_n} n_1 = \frac{1}{\pi} \int_0^{\eta_{\rm turn}} \frac{1}{\sqrt{-V(\eta)}} \frac{1}{1+\eta} d\eta$$
$$\approx \frac{1}{\pi\sqrt{-V(0)}} \int_{-1}^{1} \mathcal{G}(\eta)^2 d\eta = I_{\mathcal{G}}^0 / \left(\pi \sqrt{1 - \frac{1}{4\kappa_{eg}^2}} \right) \tag{38}$$

whereby all expressions have to be evaluated for $\alpha_n = 0$ and $\kappa = \kappa_{eg}$. We can apply the same procedure to $\partial_{\alpha_n} n_2$ and obtain

$$\partial_{\alpha_n} n_2 \approx -\frac{1}{\pi} I_{\mathcal{F}}^0 \,. \tag{39}$$

We just have to evaluate $I_{\mathcal{F}}^p$ and $I_{\mathcal{G}}^p$ for $\alpha_n = 0$ and κ_{eg} which can again be done by using equation (10) and equation (11), respectively. From now on we will denote with $I_{\mathcal{F}}^p$ and $I_{\mathcal{G}}^p$ the values for $\alpha_n = 0$ and $\kappa = \kappa_{eg}$.

4. Photon path representation

4.1. Solving the Schrödinger equation by a photon path representation

We can use the results of the previous section in order to determine the time evolution of the system. If we assume that initially only one two-level atom is in an excited state and that the field is in the vacuum state, the time evolution of the system is restricted to the subspace of the Hilbert space which corresponds one excitation only. Each state in this subspace is covered by the following ansatz for the wave function

$$|\psi(t)\rangle = b^1(t)\,|e,g\rangle^A\,|0\rangle^P + b^2(t)\,|g,e\rangle^A\,|0\rangle^P$$
$$+ \sum_i f_i(t)\,|g,g\rangle^A\,\hat{a}_i^\dagger\,|0\rangle^P \,. \tag{40}$$

(The superscripts A and P refer to atoms and photons, respectively.) The Schrödinger equation leads to a coupled system of linear differential equations. We apply the Laplace transform in order to obtain a system of linear algebraic equations. Hereby, we define the Laplace transforms of the probability amplitudes by

$$|\widetilde{\psi}(\Lambda)\rangle = \int_0^\infty e^{\mathrm{i}\Lambda t}\,|\psi(t)\rangle\,dt = \widetilde{b^1}(\Lambda)\,|e,g\rangle^A\,|0\rangle^P$$
$$+ \widetilde{b^2}(\Lambda)\,|g,e\rangle^A\,|0\rangle^P + \sum_i \widetilde{f_i}(\Lambda)\,|g,g\rangle^A\,|1\rangle_i^P \tag{41}$$

for $\mathrm{Im}(\Lambda) > 0$. By eliminating the Laplace transforms of the photonic excitations $\widetilde{f_i}(\Lambda)$, we obtain

$$\mathrm{i}\begin{pmatrix} b^1(0) \\ b^2(0) \end{pmatrix}$$
$$= \left[\Lambda - \omega_{eg} + \mathrm{i}\begin{pmatrix} A^{1,1}(\Lambda) & A^{1,2}(\Lambda) \\ A^{2,1}(\Lambda) & A^{2,2}(\Lambda) \end{pmatrix}\right] \begin{pmatrix} \widetilde{b^1}(\Lambda) \\ \widetilde{b^2}(\Lambda) \end{pmatrix} \tag{42}$$

with

$$A^{a,b}(\Lambda) = \mathrm{i}|D|^2 \sum_j \frac{\omega_j}{2\epsilon_0\hbar} \frac{(\mathbf{g}_j(\mathbf{x}_a))_z (\mathbf{g}_j(\mathbf{x}_b))_z^*}{\Lambda - \omega_j} \quad a,b \in \{1,2\} \,. \tag{43}$$

All properties of the cavity are now encoded in the functions $A^{a,b}$. It is a non trivial task to evaluate these functions because they are defined by an infinite sum over all modes j which couple to the atomic dipoles. However, by exploiting the semiclassical expressions derived previously, we are able to evaluate these functions. This way, we can solve the linear system of equations. Nevertheless, evaluating the inverse Laplace transform is still a non trivial task. For this purpose it is convenient to solve the linear system of equations by applying the Neumann series. The corresponding solution is given by

$$\begin{pmatrix} \widetilde{b}^1(\Lambda) \\ \widetilde{b}^2(\Lambda) \end{pmatrix} = \mathrm{i}\sum_{n=0}^{\infty} \frac{\left[-\mathrm{i}\begin{pmatrix} T_1^{1,1} & T_1^{1,2} \\ T_1^{2,1} & T_1^{2,2} \end{pmatrix}\right]^n}{(\Lambda + \mathrm{i}\Gamma/2 - \omega_{eg})^{n+1}} \begin{pmatrix} b^1(0) \\ b^2(0) \end{pmatrix} . \tag{44}$$

Thereby, we have defined $T_1^{a,b} = A^{a,b}(\Lambda) - \delta_{a,b}\Gamma/2$ with Γ denoting the spontaneous decay rate of the excited state $|e\rangle$. In order to enforce convergence of the series in equation (44) we can exploit the fact that for the application of the inverse Laplace transform we can choose an axis of integration with $\mathrm{Im}\,[\Lambda]$ arbitrary large. By doing so, we can make $T_1^{a,b}/(\Lambda + \mathrm{i}\Gamma/2 - \omega_{eg})$ arbitrary small and thereby enforce convergence of the series. Of course, at first sight it may appear as a major complication to apply the Neumann series in order to invert a 2 by 2 matrix but due to the retardation effects in the system caused by the finite speed of light we only have to take into account finitely many terms. However, this expansion enables us to evaluate the inverse Laplace transform and leads to a semiclassical photon path representation of the relevant probability amplitudes [15, 19]. Hereby the terms generated by the Neumann expansion in equation (44) represent sequences of spontaneous emission and absorption processes connected by the propagation of single photons. Thereby, the propagation of a photon from atom a to atom b is described by the function $T_1^{a,b}$.

4.2. Evaluating the functions $A^{a,b}(\Lambda)$

The evaluation of the functions $A^{a,b}(\Lambda)$ is the main difficulty in order to analyze the dynamics. Therefore we are going to make use of the semiclassical results obtained in Sec. 3. Due to our semiclassical quantization functions we are able to label the mode functions by the integers n_1, n_2. Therefore we obtain

$$A^{a,b}(\Lambda) = \mathrm{i}\frac{|D|^2}{2\epsilon_0\hbar} \sum_{n_1,n_2} \omega_{n_1,n_2} \frac{\left(\mathbf{g}^b_{n_1,n_2}\right)_z \left(\mathbf{g}^a_{n_1,n_2}\right)^*_z}{\Lambda - \omega_{n_1,n_2}} . \tag{45}$$

The semiclassical treatment of the mode functions delivers not only the mode functions for discrete values of n_1, n_2 but also smooth interpolations for all real numbers in between. Of course these values are unphysical, because they correspond to a violation of the quantization condition. We can however exploit this smooth interpolation by applying the Poisson summation

Journal of Physics: Conference Series **672** (2016) 012004 doi:10.1088/1742-6596/672/1/012004

formula. Thus, we obtain

$$\begin{aligned} A^{a,b}(\Lambda) &= \mathrm{i}\frac{D^2}{2\epsilon_0\hbar}\sum_{N_1,N_2\in\mathbb{Z}}\int_0^\infty\int_0^\infty \omega(n_1,n_2)\frac{\left(\mathbf{g}^a(n_1,n_2)\right)_z^*\left(\mathbf{g}^b(n_1,n_2)\right)_z}{\Lambda-\omega(n_1,n_2)}e^{\mathrm{i}2\pi(n_1N_1+n_2N_2)}dn_1dn_2 \\ &\approx \mathrm{i}\frac{D^2}{2\epsilon_0\hbar}\sum_{N_1,N_2\in\mathbb{Z}}\int_{-\infty}^\infty\int_{-\infty}^\infty \omega(n_1,n_2)\frac{\left(\mathbf{g}^a(n_1,n_2)\right)_z^*\left(\mathbf{g}^b(n_1,n_2)\right)_z}{\Lambda-\omega(n_1,n_2)}e^{\mathrm{i}2\pi(n_1N_1+n_2N_2)}dn_1dn_2 \\ &\approx \mathrm{i}\frac{D^2c_0}{d\epsilon_0\hbar}\sum_{N_1,N_2\in\mathbb{N}_0}\int_{-\infty}^\infty\int_{-\infty}^\infty \left|\mathcal{J}\begin{pmatrix} n_1(\kappa,\alpha_n) \\ n_2(\kappa,\alpha_n)\end{pmatrix}\right|_{\kappa_{eg}}\frac{\left(\mathbf{g}^a(\kappa_{eg},\alpha_n)\right)_z^*\left(\mathbf{g}^b(\kappa_{eg},\alpha_n)\right)_z}{\Lambda-2c_0\kappa/d} \\ &\cdot e^{\mathrm{i}2\pi(n_1(\kappa,\alpha_n)N_1+n_2(\kappa,\alpha_n)N_2)}d\kappa d\alpha_n \end{aligned} \tag{46}$$

with $\mathcal{J}$ denoting the Jacobian determinant. In the second and third step we have used the fact that the dynamics of the system is mainly influenced by the mode functions with $\kappa \approx \kappa_{eg}$ and $\alpha_n \approx 0$. Furthermore we have restricted the sum to nonnegative integers in the third step. A numerical calculation confirms that these terms quickly go to zero for increasing values of d/λ_{eg} and f/λ_{eg} and are already negligible for $d/\lambda_{eg}, f/\lambda_{eg} > 1$. In fact these terms are artefacts of the semiclassical approximation. Finally, we obtain

$$\begin{aligned} &A^{1,1}(\Lambda) = A^{2,2}(\Lambda) \\ &= \Gamma/2 + \sum_{(N_1,N_2)\in\mathbb{N}_0\times\mathbb{N}_0/(0,0)} e^{\mathrm{i}\tau(N_1,N_2)\Lambda}A_{N_1,N_2}\,, \end{aligned} \tag{47}$$

$$\begin{aligned} &A^{1,2}(\Lambda) = A^{2,1}(\Lambda) \\ &= \sum_{N_1\in\left[\mathbb{N}_0+\frac{1}{2}\right]}\sum_{N_2\in\mathbb{N}_0} e^{\mathrm{i}\tau(N_1,N_2)\Lambda}A_{N_1,N_2}\,, \end{aligned} \tag{48}$$

with

$$A_{N_1,N_2} = \Gamma(-1)^{N_2}\mathcal{W}(s_{\alpha_n}(N_1,N_2))\,, \tag{49}$$

(50)

$$s_{\alpha_n}(N_1,N_2) = N_1\frac{4I^0_{\mathcal{G}}}{\sqrt{1-\frac{1}{4\kappa_{eg}^2}}} - 4N_2I^0_{\mathcal{F}}\,, \tag{51}$$

$$\tau(N_1,N_2) = N_1 2d/c_0 + N_2 2f/c_0\,, \tag{52}$$

$$\mathcal{W}(x) = 3\left(-\mathrm{csch}\,(x)^2 + x\coth(x)\,\mathrm{csch}\,(x)^2\right) \tag{53}$$

and with

$$\Gamma = \Gamma_{\text{free}}\frac{I^0_{\mathcal{F}} + I^0_{\mathcal{G}}\frac{f}{d}\sqrt{1-1/(4\kappa_{eg}^2)}}{(I^2_{\mathcal{F}}I^0_{\mathcal{G}} - I^0_{\mathcal{F}}I^2_{\mathcal{G}})}^{-1}\,. \tag{54}$$

(55)

The spontaneous decay rate in free space is denoted by

$$\Gamma_{\text{free}} = \frac{D^2\omega_{eg}^3}{3c_0^3\pi\epsilon_0\hbar}\,. \tag{56}$$

Journal of Physics: Conference Series **672** (2016) 012004 doi:10.1088/1742-6596/672/1/012004

In fact the constants A_{N_1,N_2} which appear in the functions $A^{a,b}(\Lambda)$ weight the different photon paths. The corresponding time delays connected to these photon paths caused by the finite speed of light are given by $\tau(N_1, N_2)$. The expression $\Gamma/2$ which appears in equation (47) turns out to be the only expression not connected to such a time delay and describes the spontaneous decay of an excited atom. The spontaneous decay rate Γ turns out to deviate from Γ_{free} for d and f below or around λ_{eg} but quickly approaches Γ_{free} as d/λ_{eg} and f/λ_{eg} increase. In fact this deviation also turns out to be an artefact introduced by the semiclassical approximations. Thus, we replace Γ by Γ_{free} in the following. Due to the fact that the parabolic cavity is only a special case of a prolate ellipsoidal cavity we are also able to reproduce the results of Ref. [15] by considering the limit $d \to \infty$, $f = \text{const.}$.

4.3. Photon path representation in the limit of short wavelength $\lambda_{eg} \ll g, f$
In this subsection we are going to investigate the short wavelength limit $\lambda_{eg} \to 0$ with $f = \text{const.}$ and $d = \text{const.}$. In this limit most terms of the series expansion of the function $A^{a,b}(\Lambda)$ in equation (47) and equation (48) vanish and only the expressions with $2N_1 = N_2$ contribute. This directly leads to the delay times $N_2\tau$ with $\tau = (2f + d)/c_0$ which we would have expected by applying the multidimensional JWKB method [20] which is directly connected to the framework of geometrical optics. All the expressions with $2N_1 \neq N_2$ describe diffraction effects which are only relevant in case of f or d being of the order of a few wavelengths.

Additionally, we observe that in the mentioned limit the contributing terms are weighted differently. The corresponding weighting factor is given by

$$|A_{N_1,N_2}|/\Gamma = \delta_{N_2,2N_1} 3^{\frac{N_2 \log\left(\frac{\epsilon+1}{1-\epsilon}\right) \coth\left(N_2 \log\left(\frac{\epsilon+1}{1-\epsilon}\right)\right) - 1}{\sinh^2\left(N_2 \log\left(\frac{\epsilon+1}{1-\epsilon}\right)\right)}} \tag{57}$$

with $\epsilon = d/(d + 2f)$ denoting the eccentricity of the cavity which is a measure of the deviation from a sphere ($\epsilon = 0$ corresponds to a sphere and $\epsilon \to 1$ to a parabola). The coupling between the atoms increases for $\epsilon \to 0$ (which corresponds to an almost spherical symmetric cavity) and decreases for $\epsilon \to 1$. This reduction of the coupling efficiency for increasing ϵ is caused by a confinement of the wave packets, carrying the excitation from one atom to the other, to the symmetry axis of the cavity. Therefore for ϵ close to unity the electric field at the position of the second atom is almost perpendicular to the symmetry axis and thus it's coupling to the dipole matrix element of the second atom is reduced.

It is indeed possible to reproduce all the results we have obtained for the short wavelength limit including the weighting of the different photon paths also by using the multidimensional JWKB method [20] and the framework of geometrical optics.

5. Results
We start our discussion of the dynamics of the atoms by investigating the extreme multimode scenario which corresponds to $\tau\Gamma \gg 1$. In this scenario the spontaneous decay rate Γ is large compared to the frequency distance between two neighboring modes. Thus the atoms couple to many modes simultaneously. This leads to a situation which substantially deviates from the situation in the single mode regime. The typical dynamics is illustrated in figure 2 a and figure 2 b. In these two examples, we chose situations with relatively small wavelength $\lambda_{eg} \ll d, f$ in which the multidimensional JWKB-method leads to a good description of the system.

In these cases the excitation probabilities of the atoms are sharply peaked in time. After the decay of the first atom the second atom is excited after time τ, which is the typical time the photon needs to travel from the first atom to the second one as expected from geometrical optics. The remaining peaks in figure 2 a and figure 2 b can be understood in terms of descriptive

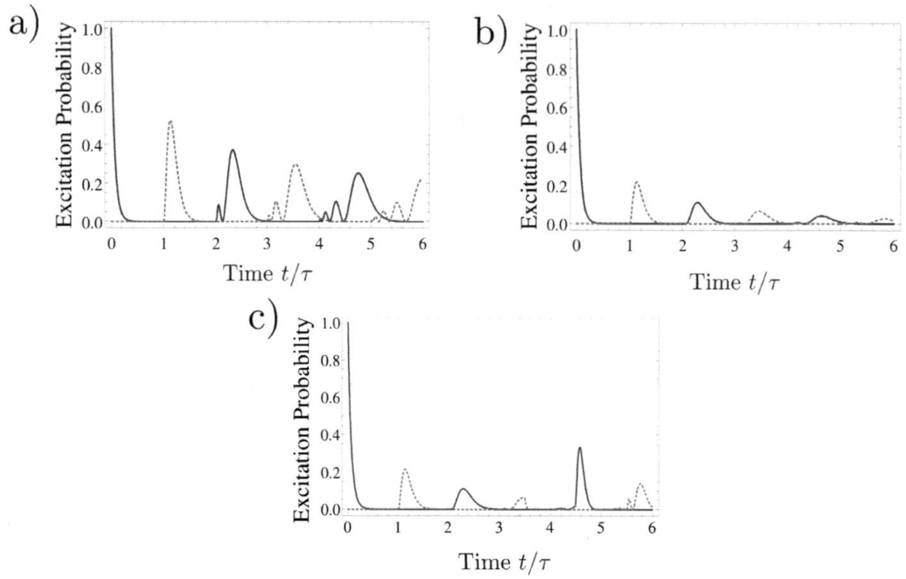

Figure 2. Atomic excitation probabilities for $\Gamma\tau = 16$ and $\epsilon = 1/10$, $\lambda_{eg} \ll d, f$ (a) and $\epsilon = 1/2$, $\lambda_{eg} \ll d, f$ (b), $d/\lambda_{eg} = 20$, $f/\lambda_{eg} = 10$ (c). The excitation probability of atom 1 (2) is represented by the blue (red) curve.

photon paths as discussed in Sec. 4. By comparing the dynamics illustrated in figure 2 a with the dynamics illustrated in figure 2 b we can study the influence of the geometry of the cavity as already discussed in Subsec. 4.3. The situation in figure 2 a with $f/d = 9/2 \Leftrightarrow \epsilon = 1/10$ corresponds to an almost spherical cavity. Therefore, the coupling between both atoms is relatively large. The situation in figure 2 b with$f/d = 1/2 \Leftrightarrow \epsilon = 1/2$ corresponds to a highly non spherical cavity. Thus, in accordance with the results obtained in Subsec. 4.3, the coupling between the atoms is reduced.

The semiclassical expressions obtained by the separation ansatz in Sec. 3 can also be used to study diffraction effects in cases of relatively long wavelengths (λ_{eg} of similar order of magnitude as d or f). Such a situation is illustrated in figure 2 c. The influence of diffraction effects can be illustrated by comparing figure 2 c and figure 2 b because both figures correspond to a cavity with eccentricity $\epsilon = 1/2$. Due to these diffraction effects, the dynamics of the system in the regime $\lambda_{eg} \approx f, g$ is much more complicated than in the regime $\lambda_{eg} \ll f, g$. In the photon path representation those diffraction effects are connected with the appearance of additional photon paths which are not connected to geometrical optics.

Our former treatment of the extreme multimode scenario also enables us to study the transition from the extreme multimode scenario which corresponds to $\tau\Gamma \gg 1$ to the single mode regime with $\tau\Gamma \ll 1$. To keep the discussion as simple as possible, we consider the scenario of very small wavelengths and an almost spherical cavity $\lambda_{eg} \ll d \ll f$. If we take the limit $\tau\Gamma \to 0$ with $\Gamma = \tau\Omega_0^2/2$ and $e^{i2\tau\omega_{eg}} = 1$, we obtain exactly the results of a single mode coupling on resonance to the two two level atoms and Ω_0 being the vacuum Rabi frequency of a single atom coupling to this mode. In figure 3 we compare the single mode scenario for $\tau \to 0$ with the results for finite τ which are more complicated due to the presence of additional modes.

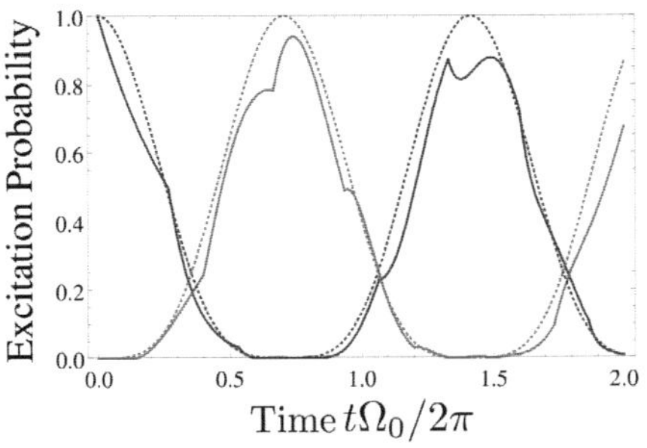

Figure 3. Comparison of atomic excitation probabilities for finite τ with ideal single mode scenario $\tau \to 0$: The dashed lines correspond to $\tau \to 0$ and the solid lines to the results for finite τ. The excitation probability of atom 1 (2) is represented by the blue (red) curves. The parameters for the solid lines are $\tau = 4\pi/15\Omega_0$.

6. Conclusion

We have investigated the dynamics of spontaneous photon emission and absorption processes of two two-level atoms trapped close to the focal points of a prolate-ellipsoidal cavity. Our theoretical approach is based on a full multimode treatment of the electromagnetic radiation field and incorporates the dipole approximation and the RWA. In order to deal with the electromagnetic radiation field in the multimode scenario, we have applied semiclassical methods by exploiting the separability of the relevant electromagnetic mode functions. With the help of the expressions obtained by our semiclassical treatment, we have developed a semiclassical photon path representation of the relevant probability amplitudes. This photon path representation enables us to discuss the dynamics inside the cavity by means of descriptive photon paths. This way, we have studied the interplay between both atoms mediated by the radiation field and have addressed intermediate instances between well-known quantum-optical scenarios. These well-known scenarios include Rabi oscillations which emerge for $\Gamma\tau \ll 1$ or an almost Markovian dynamics of the atoms dominated by spontaneous decay processes which emerge as $\Gamma\tau \gg 1$ and as the eccentricity ϵ approaches unity. We have also investigated diffraction effects and their suppression in the limit of extremely small wavelengths.

Acknowledgments

This work is supported by CASED III, the BMBF Project Q.com, and by the DFG as part of the CRC 1119 CROSSING.

References

[1] Berman P (ed) 1994 *Cavity Quantum Electrodynamics* (San Diego: Academic Press)
[2] Walther H H *et al.* 2006 *Prog. Phys.* **69** 1325
[3] Haroche S *et al.* 2006 *Exploring the Quantum: Atoms, Cavities and Photons* (Oxford: Oxford University Press)
[4] Goy P *et al.* 1983 *Phys. Rev. Lett.* **50** 1903–1906
[5] Meschede D *et al.* 1985 *Phys. Rev. Lett.* **54** 551
[6] McKeever J *et al.* 2003 *Nature* **425** 268–271
[7] Schleich W 2001 *Quantum Optics in Phase Space* (Berlin: Wiley-VCH)
[8] Jaynes E T and Cummings F W 1963 *Proceedings of the IEEE* **51** 89–109
[9] Moehring D L, Maunz P, Olmschenk S, Younge K C, Matsukevich D N, Duan L M and Monroe C 2007 *Nature* **449** 68
[10] Olmschenk S, Matsukevich D N, Maunz P, Hayes D, Duan L M and Monroe C 2009 *Science* **323** 486
[11] Aljunid S, Maslennikov G, Wang Y, Dao H, Scarani V and Kurtsiefer C 2013 *Phys. Rev. Lett.* **111**(10) 103001
[12] Maiwald R *et al.* 2009 *Nat. Phys.* **5** 551–554

[13] Maiwald R *et al.* 2012 *Phys. Rev. A* **86** 043431
[14] Hétet G, Slodička L, Hennrich M and Blatt R 2011 *Physical Review Letters* **107** 133002
[15] Alber G *et al.* 2013 *Phys. Rev. A* **88** 023825
[16] Flammer C 1957 *Spheroidal Wave Functions* (Stanford University Press)
[17] Berry M and Mount K 1972 *Rep. Prog. Phys.* **35** 315
[18] Abramowitz M and Stegun I 1984 *Pocketbook of Mathematical Functions* (Harri Deutsch)
[19] Milonni P W and Knight P L 1974 *Phys. Rev. A* **10** 1096–1108
[20] Maslov V P and Fedoriuk M V 1981 *Semi-Classical Approximation in Quantum Mechanics* (Dordrecht: D. Reidel)

Journal of Physics: Conference Series **672** (2016) 012005 doi:10.1088/1742-6596/672/1/012005

Radiation from a longitudinal oscillator moving along a circular trajectory

A S Kotanjyan and A A Saharian

Yerevan State University, 1 Alex Manoogian Street, 0025 Yerevan, Armenia
Institute of Applied Problems in Physics, 25 Nersessian Street, 0014 Yerevan, Armenia

E-mail: anna.kotanjyan@ysu.am, saharian@ysu.am

Abstract. Radiation intensity is investigated for a charged longitudinal oscillator moving along a helical trajectory around a dielectric cylinder immersed into a homogeneous medium. The influence of the longitudinal oscillations one on the characteristics of the angular distribution of the radiation intensity in the exterior medium is discussed. It is shown that the presence of oscillations serves as a mechanism to control the location and the height of the radiation peaks.

1. Introduction

The presence of medium can essentially change the characteristics of the high-energy electromagnetic processes. Moreover, new types of radiation phenomena can arise. Well-known examples are Cherenkov and transition radiations. In particular, by taking into account the wide applications of the synchrotron radiation, it is important to study the influence of medium on the spectral and angular characteristics of the synchrotron radiation.

In references [1]-[4] we have investigated the radiation generated by a charge moving along a helical orbit around/inside a dielectric cylinder. It has been shown that under certain conditions on the material of the cylinder and on the particle velocity, strong narrow peaks appear in the angular density of the radiation intensity. At these peaks the radiated energy exceeds the corresponding quantity in the case of a homogeneous medium by several orders of magnitude. In these investigations it was assumed that the charge moves along a circular helix coaxial with the cylinder axis and the rotation in the plane perpendicular to this axis is uniform. In realistic situations the trajectory of the particle may differ from the coaxial one. In Ref. [5] we have considered the electromagnetic fields and the radiation intensity for a charge moving around a dielectric cylinder along a helical trajectory the projection of which on the plane perpendicular to the cylinder axis is an arbitrary closed curve. In particular, the influence of the trajectory shift from the circular one on the characteristics of the peaks is discussed.

In the present paper, as an application of general results of [5], we consider the radiation from a longitudinal oscillator moving along a circular helical trajectory around a dielectric cylinder. In Section 2, the formula for the angular distribution of the radiation intensity on a given harmonic is presented for a general case of helical motion. The special case of a longitudinal oscillator moving along a circular helix is discussed in Section 3. The main results are summarized in Section 4.

Journal of Physics: Conference Series **672** (2016) 012005 doi:10.1088/1742-6596/672/1/012005

2. Radiation intensity for a general motion

Consider a point charge q moving along a helical trajectory around a dielectric cylinder with radius ρ_c. The dielectric permittivity of the cylinder will be denoted by ε_0 and the cylinder is immersed into a homogeneous medium with permittivity ε_1. In cylindrical coordinates (ρ, ϕ, z), with z-axis along the cylinder axis, the motion of the charge is described by the functions $\rho = \rho_e(t)$, $\phi = \phi_e(t)$, $z = v_{\|}t$. We will denote by T the period of the transverse motion and $\omega_0 = 2\pi/T$. Introducing the angle θ between z-axis and the radiation direction, at large distances from the cylinder, for the angular density of the radiated energy per unit time on a given harmonic $n = 1, 2, \ldots$, one has

$$\frac{dI_n}{d\Omega} = \frac{q^2}{2\pi^3 c}\frac{n^2\omega_0^2\sqrt{\varepsilon_1}}{\left|1-\beta_{1\|}\cos\theta\right|^3}\sum_{m=-\infty}^{\infty}\left[\left|D_{m,n}^{(+1)} - D_{m,n}^{(-1)}\right|^2 + \left|D_{m,n}^{(+1)} + D_{m,n}^{(-1)}\right|^2\cos^2\theta\right], \tag{1}$$

with the notation $\beta_{1\|} = v_{\|}\sqrt{\varepsilon_1}/c$. Here, we have defined

$$\begin{aligned} D_{m,n}^{(p)} &= \frac{\pi}{2ic}\left[J_{m+p,m}(n,\lambda_1) - H_{m+p,m}(n,\lambda_1)\frac{W(J_{m+p},J_{m+p})}{W(J_{m+p},H_{m+p})}\right] \\ &+\frac{i\pi\lambda_1}{2ck_z}\left[J_{m,m,z}(n,\lambda_1) - H_{m,m,z}(n,\lambda_1)\frac{W(J_m,J_m)}{W(J_m,H_m)}\right] \\ &+\frac{pJ_m(\lambda_0\rho_c)}{c\rho_c\alpha_m}\frac{J_{m+p}(\lambda_0\rho_c)}{W(J_{m+p},H_{m+p})}\left[\frac{k_{zn}H_{m,m,z}(n,\lambda_1)}{W(J_m,H_m)} + \frac{\lambda_0}{2}\sum_{l=\pm1}\frac{H_{m+l,m}(n,\lambda_1)}{W(J_{m+l},H_{m+l})}\right], \end{aligned} \tag{2}$$

where $J_m(x)$ and $H_m(x) = H_m^{(1)}(x)$ are the Bessel function and the Hankel function of the first kind,

$$\begin{aligned} \lambda_1 &= \frac{\omega_n}{c}\sqrt{\varepsilon_1}\sin\theta,\ \lambda_0 = \frac{\omega_n}{c}\sqrt{\varepsilon_0 - \varepsilon_1\cos^2\theta}, \\ \omega_n &= \frac{n\omega_0}{1-\beta_{1\|}\cos\theta},\ k_{zn} = \frac{\omega_n}{c}\sqrt{\varepsilon_1}\cos\theta, \end{aligned} \tag{3}$$

and

$$\alpha_m = \frac{\varepsilon_0}{\varepsilon_1 - \varepsilon_0} - \frac{\lambda_0 J_m(\lambda_0\rho_1)}{2}\sum_{l=\pm1} l\frac{H_{m+l}(\lambda_1\rho_1)}{W(J_{m+l},H_{m+l})}. \tag{4}$$

Other functions are defined in accordance with

$$W(J_{m+p},F_{m+p}) = J_m(\lambda_0\rho_1)\frac{\partial F_m(\lambda_1\rho_1)}{\partial\rho_1} - F_m(\lambda_1\rho_1)\frac{\partial J_m(\lambda_0\rho_1)}{\partial\rho_1}, \tag{5}$$

where $F = J, H$, and

$$\begin{aligned} F_{m',m,l}(n,\lambda_1) &= \frac{1}{T}\int_0^T dt\, v_l(t)F_{m'}(\lambda_1\rho_e(t))e^{-im\phi_e(t)+in\omega_0 t}, \\ F_{m+p,m}(n,\lambda_1) &\equiv F_{m+p,m,\phi}(n,\lambda_1) - ipF_{m+p,m,\rho}(n,\lambda_1), \end{aligned} \tag{6}$$

with $l = \rho, \phi, z$, $v_z = v_{\|}$. For a given direction, the radiation frequency is given by $|\omega_n|$. For $\beta_{1\|} > 1$, the radiation along the Cherenkov direction $\theta = \theta_C = \arccos(c/(v_{\|}\sqrt{\varepsilon_1}))$ should be considered separately. For a charge rotating near the boundary of the dielectric waveguide, for the wavelength of the radiated quanta one gets

$$\lambda_{(n)} \approx \frac{2\pi\rho_c}{n\beta_{1\perp}}|1-\beta_{1\|}\cos\theta|,\ \beta_{1\perp} = v_{1\perp}\sqrt{\varepsilon_1}/c. \tag{7}$$

If $\beta_{1\|}\cos\theta$ is not too close to 1, for the waveguide radius $\rho_c \sim 1$cm and for harmonics $n \sim 100$ the frequency of the radiated wave is in the terahertz range. By decreasing the radius of the waveguide, the radiation frequency increases. For $\rho_0 \sim 0.1$mm the frequency is near the optical range. Currently dielectric waveguides are available with the radius of the order of nanometer (see, for instance, [6, 7]).

In a special case of a circular helix with $\rho_e(t) = \rho_e = \text{const}$, $\phi_e(t) = \omega_0 t$, one has $F_{m',m,l}(n,\lambda_1) = v_l F_{m'}(\lambda_1\rho_e)\delta_{nm}$ and in (1) the term $m = n$ contributes only. The radiation intensity in this special case has been discussed in detail in [4]. In particular, the conditions were specified under which strong narrow peaks appear in the angular distribution for the radiation intensity at a given harmonic n.

3. Radiation intensity from a longitudinal oscillator

In this section we specify the general formula for the radiation intensity given above for a special case of the motion along a circular helix with additional longitudinal oscillations of the charge. In this case the components of the velocity are given by the expressions

$$\begin{aligned} v_\rho(t) &= 0,\ v_z(t) = v_\|, \\ v_\phi(t) &= v_\perp[1 + a\cos(n_0\omega_0 t)],\ v_\perp = \omega_0\rho_e, \end{aligned} \tag{8}$$

with n_0 being an integer. For the angular velocity of the rotation and for the angular coordinate one has

$$\begin{aligned} \Omega_\phi(t) &= \omega_0[1 + a\cos(n_0\omega_0 t)], \\ \phi_e(t) &= \omega_0[t + \frac{a}{n_0\omega_0}\sin(n_0\omega_0 t)]. \end{aligned} \tag{9}$$

For the motion (8), by taking unto account that $\rho_e(t) = \rho_e = \text{const}$ and $v_\rho = 0$, we see that $F_{m+p,m,\rho}(n,\lambda_1) = 0$. For the remaining functions, with the help of the formula

$$e^{i\alpha\sin\xi} = \sum_{s=-\infty}^{+\infty} J_s(\alpha)e^{is\xi}, \tag{10}$$

for $m \neq 0$ one gets the following expressions

$$\begin{aligned} F_{m+p,m,\phi}(n,\lambda_1) &= F_{m+p,m}(n,\lambda_1) = \frac{n}{m}v_\perp F_{m+p}(\lambda_1\rho_e)J_{\frac{n-m}{n_0}}\left(\frac{ma}{n_0}\right), \\ F_{m+p,m,z}(n,\lambda_1) &= v_\| F_{m+p}(\lambda_1\rho_e)J_{\frac{n-m}{n_0}}\left(\frac{ma}{n_0}\right). \end{aligned} \tag{11}$$

In these expressions $(n-m)/n_0$ should be an integer which we will denote by s. Hence, for the allowed values of m one has

$$m = n - sn_0,\ s = 0, \pm1, \pm2, \ldots \tag{12}$$

So, for $m \neq 0$ we get

$$D_{m,n}^{(p)} = J_s\left(\frac{ma}{n_0}\right)\overline{D}_{m,n}^{(p)}, \tag{13}$$

where

$$\begin{aligned}\overline{D}^{(p)}_{m,n} &= \frac{\pi v_\perp n}{2icm}\left[J_{m+p}(\lambda_1\rho_e) - H_{m+p}(\lambda_1\rho_e)\frac{W(J_{m+p},J_{m+p})}{W(J_{m+p},H_{m+p})}\right] \\ &+\frac{i\pi v_\parallel \lambda_1}{2ck_z}\left[J_m(\lambda_1\rho_e) - H_m(\lambda_1\rho_e)\frac{W(J_m,J_m)}{W(J_m,H_m)}\right] + \frac{pJ_m(\lambda_0\rho_c)}{\rho_c\alpha_m}\frac{J_{m+p}(\lambda_0\rho_c)}{W(J_{m+p},H_{m+p})} \\ &\times\left[\frac{k_z H_m(\lambda_1\rho_e)}{W(J_m,H_m)}\frac{v_\parallel}{c} + \frac{\lambda_0}{2}\frac{n}{m}\frac{v_\perp}{c}\sum_{l=\pm1}\frac{H_{m+l}(\lambda_1\rho_e)}{W(J_{m+p},H_{m+p})}\right]. \end{aligned} \tag{14}$$

The expression for $\overline{D}^{(p)}_{m,n}$ coincides with the expression of $D^{(p)}_{m,n}$ in the case of a uniform circular motion with $v_\phi(t) = \omega_0\rho_e$ if we replace $v_\perp \to v_\perp n/m$.

The term $m = 0$ contributes for $n = n_0$ only. In this case we have $F_{m+p,m,z}(n,\lambda_1) = 0$ and

$$F_{m+p,m}(n,\lambda_1) = F_{m+p,m,\phi}(n,\lambda_1) = \frac{v_\perp}{2}aF_p(\lambda_1\rho_e)\delta_{n,n_0}. \tag{15}$$

For the functions $D^{(p)}_{m,n}$ one finds

$$D^{(1)}_{0,n} = -D^{(-1)}_{0,n} = \frac{\pi a v_\perp}{4ic}\left[J_1(\lambda_1\rho_e) - H_1(\lambda_1\rho_e)\frac{W(J_1,J_1)}{W(J_1,H_1)}\right]. \tag{16}$$

Hence, for the contribution of the $m = 0$ term, which is present only for $n = n_0$, one gets

$$\frac{dI_{n,m=0}}{d\Omega} = \frac{q^2v_\perp^2a^2}{4\pi c^3}\frac{n^2\omega_0^2\sqrt{\varepsilon_1}}{\left|1-\beta_{1\parallel}\cos\theta\right|^3}\left|J_1(\lambda_1\rho_e) - H_1(\lambda_1\rho_e)\frac{W(J_1,J_1)}{W(J_1,H_1)}\right|^2. \tag{17}$$

In the case $n \neq n_0$, for the angular density of the radiation intensity on a given harmonic n we find the expression

$$\frac{dI_n}{d\Omega} = \frac{q^2}{2\pi^3 c}\frac{n^2\omega_0^2\sqrt{\varepsilon_1}}{\left|1-\beta_{1\parallel}\cos\theta\right|^3}\sum_{s=-\infty}^{+\infty}{}' J_s^2(\frac{ma}{n})\left[\left|\overline{D}^{(+1)}_{m,n} - \overline{D}^{(-1)}_{m,n}\right|^2 + \left|\overline{D}^{(+1)}_{m,n} + \overline{D}^{(-1)}_{m,n}\right|^2\cos^2\theta\right], \tag{18}$$

where m is given by (12) and the prime means that the term with $s = n/n_0$ should be excluded from the summation. For $a \ll 1$ the dominant contribution comes from the term with $s = 0$ and the contributions of the other terms are small by the factor a^{2s}. The leading term coincides with the corresponding quantity for a charge uniformly rotating around a dielectric cylinder. For $n = n_0$ the radiation intensity is the sum of the expressions given by (17) and (18):

$$\begin{aligned}\frac{dI_{n_0}}{d\Omega} &= \frac{q^2}{2\pi^3 c}\frac{n_0^2\omega_0^2\sqrt{\varepsilon_1}}{\left|1-\beta_{1\parallel}\cos\theta\right|^3}\left\{\frac{\pi^2v_\perp^2a^2}{2c^2}\left|J_1(\lambda_1\rho_e) - H_1(\lambda_1\rho_e)\frac{W(J_1,J_1)}{W(J_1,H_1)}\right|^2\right. \\ &+\left.\sum_{s=-\infty,\neq0}^{+\infty} J_{s-1}^2(sa)\left[\left|\overline{D}^{(+1)}_{sn_0,n_0} - \overline{D}^{(-1)}_{sn_0,n_0}\right|^2 + \left|\overline{D}^{(+1)}_{sn_0,n_0} + \overline{D}^{(-1)}_{sn_0,n_0}\right|^2\cos^2\theta\right]\right\}. \end{aligned} \tag{19}$$

In figures below we have presented the angular density of the number of the radiated quanta, per period T of the transversal motion,

$$\frac{dN_n}{d\Omega} = \frac{T}{\hbar\left|\omega_n\right|}\frac{dI_n}{d\Omega}, \tag{20}$$

as a function of the angle θ. The graphs are plotted for the harmonic $n = 10$. The dashed curves correspond to the radiation in a homogeneous medium with permittivity ε_1 and the full curves correspond to the radiation from a charge rotating around a dielectric cylinder with permittivity $\varepsilon_0 = 3.75\varepsilon_1$ immersed into a homogeneous medium with permittivity ε_1 (for a charge rotating in the vacuum ($\varepsilon_1 = 1$), $\varepsilon_0 = 3.75$ corresponds to the dielectric permittivity of fuzed quartz). In all cases we have taken $\rho_c/\rho_e = 0.95$. For figure 1 one has $\beta_{1\|} = 0.35$ and $\beta_{1\perp} = 0.9$. The left panel corresponds to the radiation in the absence of the longitudinal oscillations ($a = 0$) and for the right panel $a = 0.1$. Figure 2 presents the same graphs for a purely transverse motion with the same energy, $\beta_{1\|} = 0$, $\beta_{1\perp} = \sqrt{0.35^2 + 0.9^2} \approx 0.966$. In this case the angular density is symmetric with respect to $\theta = \pi/2$. For the angular locations and heights of the left two peaks on the right panel of figure 2, induced by longitudinal oscillations, one has $(0.52, 19.35)$ and $(0.76, 556.2)$. For the left panel of figure 3 we have taken $a = 0.2$ and for the right panel $n_0 = 3$. For both panels the values of the other parameters are the same as those on the right panel of figure 1.

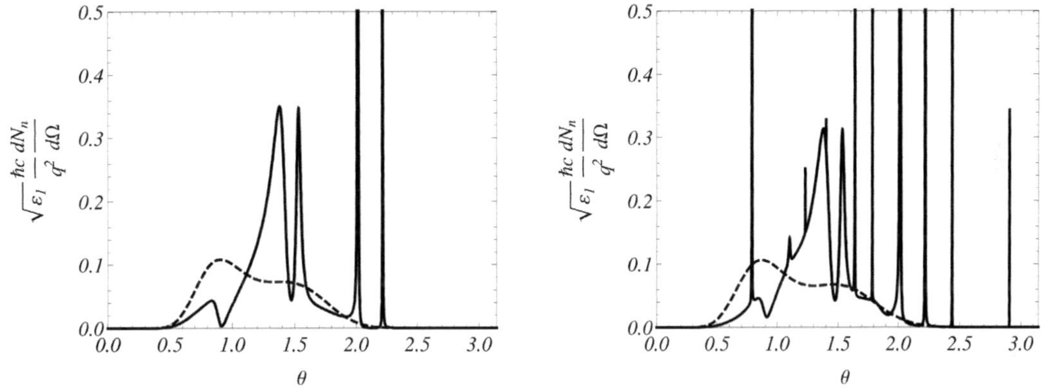

Figure 1. The angular density of the number of the quanta, radiated per period of the transverse motion on the harmonic $n = 10$, as a function of the angle θ. The dashed curves correspond to the radiation in a homogeneous medium with dielectric permittivity ε_1 and the full curves correspond to the radiation from a charge rotating around a dielectric cylinder with permittivity ε_0 immersed in the same medium. The left panel corresponds to the radiation in the absence of the longitudinal oscillations $a = 0$ and for the right panel $a = 0.1$, $n_0 = 2$. For the values of the other parameters we have taken $\rho_c/\rho_e = 0.95$, $\beta_{1\|} = 0.35$, $\beta_{1\perp} = 0.9$.

As it is seen from the graphs, the presence of longitudinal oscillations leads to the appearance of new peaks in the angular density of the radiated quanta. The number of peaks is increasing with the increase of the amplitude and frequency of the longitudinal oscillations.

4. Conclusion

We have investigated the radiation intensity of a charged longitudinal oscillator moving along an helical trajectory around a dielectric cylinder. Similar to the case of coaxial circular motion under certain conditions for the parameters of the trajectory and dielectric cylinder strong narrow peaks appear in the angular distribution of the radiation intensity in the exterior medium. Instead of a single peak in the case of a uniform coaxial circular motion, for an longitudinal oscillator set of peaks appear. The increase of the oscillating amplitude leads to the increase of the number of the peaks and the peaks are shifted to the direction of small angles. Heights of the peaks may either decrease or increase compared with the case of the uniform motion. The frequency

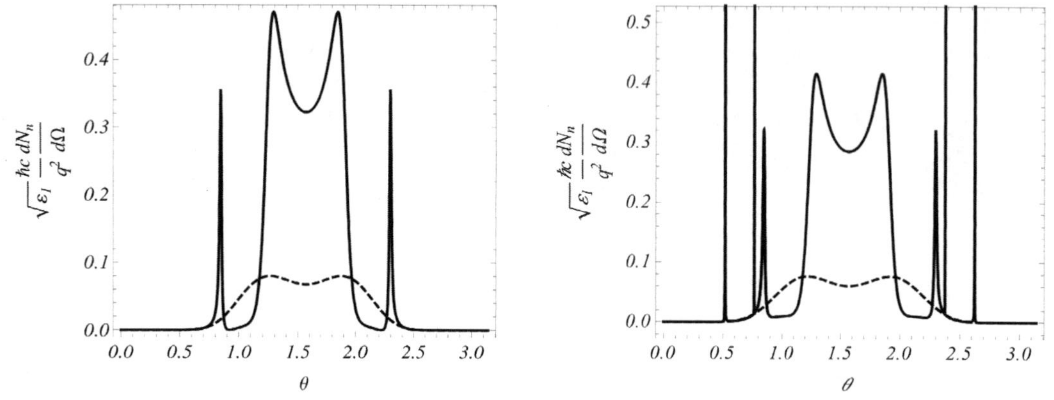

Figure 2. The same graphs as in figure 1 for a purely transverse motion: $\beta_{1\|} = 0$, $\beta_{1\perp} \approx 0.966$.

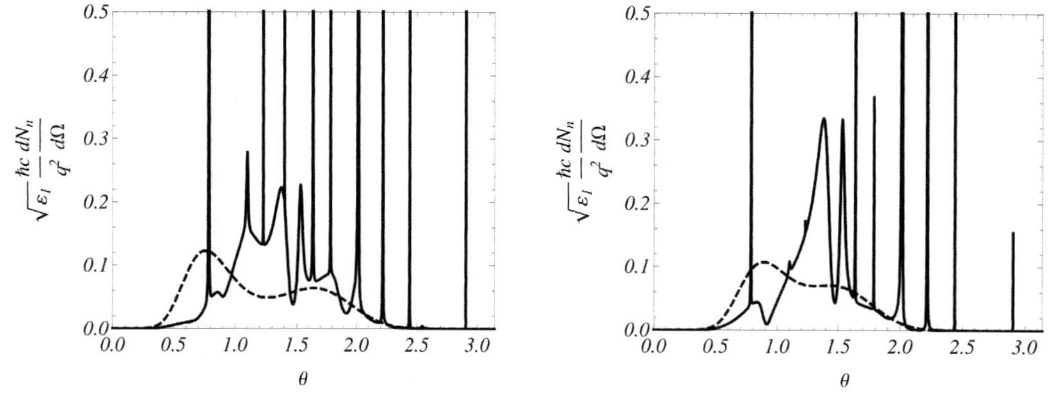

Figure 3. The same as in the right panel of figure 1. For the left panel $a = 0.2$, $n_0 = 2$ and for the right panel $a_0 = 0.1$, $n_0 = 3$. For both panels the values of the other parameters are the same as those on the right panel of figure 1.

of the emitted quanta at the peaks depends on the angular location of the peak. In the case of spiral motion the frequency increases due to the Doppler shift.

References

[1] Saharian A A Kotanjyan A S and Grigoryan M L 2007 *J. Phys. A* **40** 1405
[2] Saharian A A and Kotanjyan A S 2007 *J. Phys. A* **40** 10641
[3] Arzumanyan S R Grigoryan L Sh Khachatryan H F Kotanjyan A S and Saharian A A 2008 *Nucl. Instrum. Methods* **B266** 3703
[4] Saharian A A and Kotanjyan A S 2009 *J. Phys. A* **42** 135402
[5] Kotanjyan A S and Saharian A A 2014 *J. Phys.: Conf. Ser.* **517** 012025
[6] Balykin V I Hakuta K Kien Fam Le *et. al.* 2004 *Phys. Rev. A* **70** 011401
[7] Afanasiev A and Minogin V 2010 *Phys. Rev. A* **82** 052903

Journal of Physics: Conference Series **672** (2016) 012006 doi:10.1088/1742-6596/672/1/012006

Effect of optomechanical coupling on squeezed-spin states

N Aggarwal[1], B Joshi[2] and A B Bhattacherjee[3]

[1]Department of Physics and Astrophysics, University of Delhi, Delhi-110007, India
[2]Department of Science & Technology, Technology Bhavan, New Mehrauli Road, New Delhi – 110016, India
[3]School of Physical Sciences, Jawaharlal Nehru University, New Delhi-110067, India

E-mail: bhattach@mail.jnu.ac.in

Abstract. We investigate the effect of optomechanical coupling on the squeezed-spin states for a Bose-Einstein Condensate embedded within the lossless optomechanical cavity for the three special cases of initial states of cavity field, namely, a coherent state, a squeezed vacuum state and a squeezed state. We show that the radiation pressure or pondermotive force acting on the cavity end mirror plays a significant role in producing the atomic-squeezed states by producing squeezed states of the cavity field which is then transferred to the condensate. We further show that the maximum spin-squeezing along the x-direction is obtained in the presence of optomechanical coupling for the initial cavity field prepared in the amplitude squeezed state, whereas, squeezing along the y-direction reaches a maximum value in the absence of optomechanical coupling for the initial coherent cavity field. We also study the additional effect of nonlinear atomic interaction on spin-squeezing.

1. Introduction

Spin squeezing [1–3] has attracted considerable attention in both the theoretical and experimental zones for over a decade. In order to improve the measurements precision in experiments [2–5] and to study the correlations and entanglement between particles [6–8], spin squeezed states are considered to be a very useful quantum resource. The phenomenon of spin-squeezing in collective spin system has also grabbed great interest due to its application in atomic clocks for reducing quantum noise [2, 3, 9] and quantum information [6, 10]. Note that different definitions of spin squeezing can be used and the most widely used spin squeezing parameters were proposed by Kitagawa and Ueda in [1] and by Wineland in [2, 3]. For decades, squeezed states of electromagnetic field have also been the very popular topic of research [11–13]. It can be produced by exploiting the optomechanical coupling between the light and movable mirror, which is also termed as ponderomotive squeezing [14]. Pondermotively generated squeezed light has been studied both theoretically [14] and experimentally [15,16]. Furthermore, spin squeezing in atomic ensembles can be produced via interaction between light and the atom, which involves the transferring of squeezing from light to atoms [3, 17]. The production of atomic-squeezed states in a two-component Bose-Einstein Condensates (BECs) via nonlinear atom-atom interaction has also been studied [6, 18, 19]. The spin-squeezed states in a BEC can be experimentally realized [18,20], which can be used to detect weak forces [21]. Motivated by these interesting developments in this field, we propose an optomechanical system consisting of an elongated cigar-shaped gas of two-level BEC atoms interacting with a single mode of a lossless optical cavity with one movable mirror. We study this model to generate the atomic-squeezed states in both the x- and y- directions for the three special cases of initial states of cavity field, namely, a coherent state, a squeezed vacuum state and a squeezed state. We further discuss how the pondermotive force acting on the end mirror as a result of optomechanical coupling helps in generating atomic-squeezed states and how this squeezing can be coherently controlled by the light-mirror coupling. We also show the additional effect of nonlinear atom-atom interaction on spin-squeezing.

2. The Model

We consider an optomechanical system consisting of a lossless Fabry-Perot optical cavity with one heavy fixed mirror and another highly reflecting end mirror which is light and allowed to move freely, as shown in figure 1. In addition, we have an elongated cigar-shaped gas of N BEC atoms of ^{87}Rb having two different hyperfine levels $|F=1, m_f=-1>$ and $|F=2, m_f=1>$ with mass m and transition frequency ω_a. In the two-mode approximation, each atomic mode is associated with an annihilation operator c_j ($j=1,2$). The cloud of BEC atoms interacts with a single quantized optical cavity mode having frequency ω_c of an optical cavity of length L. A single-mode quantized cavity field is equivalent to a single quantum-mechanical harmonic oscillator of unit mass. The external pump laser drives the optomechanical system through the fixed cavity mirror. Note that the movable mirror is considered as a single mechanical oscillator of frequency ω_m and mass M_m, which can be realized experimentally by using a bandpass filter in the detection loop such that the other mechanical degrees of freedom arising from the radiation pressure can be neglected [22]. The simplest model of such an optomechanical system is provided by the following Hamiltonian in the dipole approximation ($\hbar=1$ throughout the paper) [23,24]:

$$H=\omega_a J_z+\omega_c a^\dagger a+\omega_m b^\dagger b+\frac{\gamma}{N}J_z^2+\frac{g_0}{\sqrt{N}}(a+a^\dagger)J_x+\varepsilon a^\dagger a(b+b^\dagger), \tag{1}$$

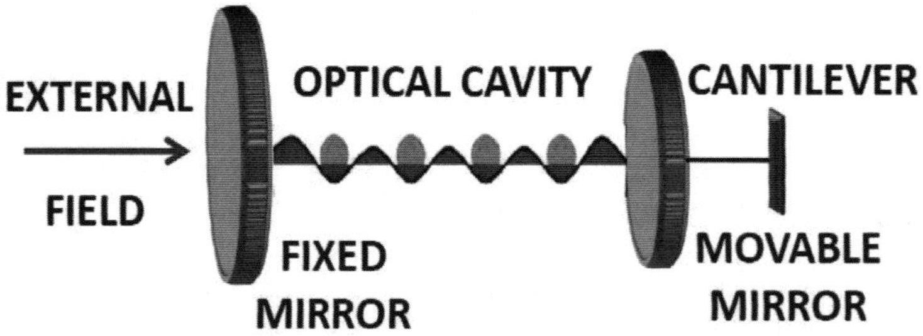

Figure 1. (color online) Schematic representation of the system involving Bose-Einstein Condensate inside a lossless optical cavity driven by an external field. Here, one of the cavity mirrors is movable.

The ensemble of N atoms is described using the picture of a collective spin operators such that $J_x=(c_1^\dagger c_2+c_2^\dagger c_1)/2$, $J_y=(c_1^\dagger c_2-c_2^\dagger c_1)/2i$ and $J_z=(c_1^\dagger c_1-c_2^\dagger c_2)/2$ [25], satisfying the commutation relations $[J_+,J_-]=2J_z$ and $[J_\pm,J_z]=\mp J_\pm$ where $J_+=c_1^\dagger c_2$ and $J_-=c_2^\dagger c_1$. The Hilbert space is spanned by symmetric Dicke states $|J,M>$ where $M=-J,-J+1.......J-1,J$ with the total spin length $J=N/2$ [26]. Moreover, the operators act on these states as $J_\pm|J,M>=\sqrt{J(J+1)-M(M\pm1)}|J,M\pm1>$, $J_z|J,M>=M|J,M>$ and

$J^2 |J,M\rangle = J(J+1)|J,M\rangle$. The lowering (raising) operator is given by a ($a^\dagger$) such that $[a,a^\dagger]=1$. The annihilation and creation operators for the mechanical oscillator are denoted by b ($b^\dagger$) satisfying the commutation relation $[b,b^\dagger]=1$. The effective nonlinear atomic interaction strength and atom-photon coupling strength are represented by γ and g_0 respectively. Origin of last term in the Hamiltonian is the result of radiation pressure exerted by the intra-cavity photons on the mirror such that the nonlinear dispersive coupling between the cavity field and the oscillating mirror is denoted by ε.

The optomechanical Hamiltonian given by equation (1) can be rewritten in the interaction picture as [27]:

$$H = \frac{\gamma}{N} J_z^2 + \varepsilon a^\dagger a (b e^{-i\omega_m t} + b^\dagger e^{i\omega_m t}) + \frac{g_0 a}{2\sqrt{N}} [J_+ e^{i(\omega_a-\omega_c)t} + J_- e^{-i(\omega_a+\omega_c)t}] + \frac{g_0 a^\dagger}{2\sqrt{N}} [J_+ e^{i(\omega_a+\omega_c)t} + J_- e^{-i(\omega_a-\omega_c)t}] \quad (2)$$

Now, we treat the mechanical mode operator b semiclassically so that it can be replaced by its real scalar quantity β with $|\beta| \gg 1$. Therefore, for $\omega_a = \omega_c$, the resulting simplified Hamiltonian in the rotating wave approximation becomes:

$$H = \frac{\gamma}{N} J_z^2 + g(t) a^\dagger a + \frac{g_0}{2\sqrt{N}} [a J_+ + a^\dagger J_-] \quad (3)$$

where $g(t) = 2\varepsilon\beta\cos(\omega_m t)$ is the modified mirror-photon coupling, which is related to the time modulated frequency of the cavity arising due to the motion of the cavity boundary [28]. In the next section, we study the generation of spin squeezing when an ensemble of BEC atoms held within the optomechanical cavity interacts with the single-mode quantized cavity field (harmonic oscillator) through the interaction Hamiltonian (3). We first focus our attention in studying the effect of pondermotive force on spin-squeezing, i.e., $\gamma = 0$. Later we discuss the effect of atom-atom interaction (γ) on atomic-squeezing.

3. Preparation of atomic-squeezed states

Here, we study the preparation of spin-squeezed states for some special cases of initial cavity field by numerically solving the Schrodinger's equation for the interaction Hamiltonian. To assess the effect of pondermotive squeezing on the generation of spin-squeezing, we first discuss the situation of an initially coherent intracavity field and then that of an initially squeezed vacuum and squeezed cavity field. The collective spins is assumed to be initially prepared in the Dicke state $|J, M=-J\rangle$ with $M = \langle J_z \rangle$. Also, $\langle J_x(0)\rangle = \langle J_y\rangle = 0$, $\Delta J_x(0) = \Delta J_y(0) = \sqrt{J/2}$ and $\Delta J_z(0) = 0$. The square root of the variance of any operator A is denoted by $\Delta A = \sqrt{\langle A^2\rangle - \langle A\rangle^2}$. The position and momentum quadratures of the single-mode quantized cavity field (harmonic oscillator) can be written in terms of cavity field operators a and $a^\dagger$ as $x(t) = [a(t) + a^\dagger(t)]/\sqrt{2}$ and $p(t) = [a(t) - a^\dagger(t)]/i\sqrt{2}$ respectively. We can characterize the squeezing of the harmonic oscillator by the parameters $\chi_x(t) = \Delta x(t)/x_0$ or $\chi_p(t) = \Delta p(t)/x_0$, with x_0 [$x_0 = \Delta x$ (coherent state)] denoting the zero-point amplitude. The condition $\chi_x < 1$ represents amplitude squeezing and $\chi_p < 1$ represents momentum squeezing.

The squeezing parameter that we are considering here is given by Kitagawa and Ueda [1] and is defined as $\zeta_S(t) = (\sqrt{2/J})\Delta J_\perp$, where the subscript $\perp$ refers to an axis perpendicular to the mean angular momentum $<J>$ where the minimum value of ΔJ is obtained. The parameter $\zeta_S(t)$ indicates the degree of quantum correlations among the elementary spins and $\zeta_S < 1$ signifies atomic squeezing. In this view, the squeezing parameters considered are defined as follows:

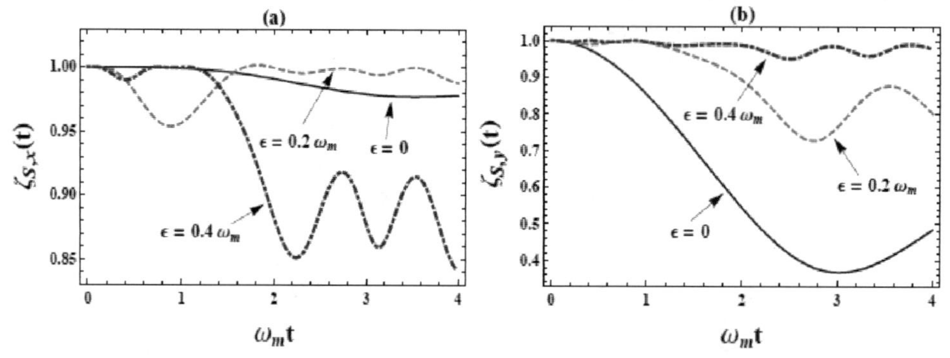

Figure 2. (color online) Squeezing parameter $\zeta_S(t)$ versus scaled time ($\omega_m t$) in absence of cavity-mirror coupling $\varepsilon = 0$ (solid line) and for two different values of mirror-photon coupling with $\varepsilon = 0.2\omega_m$ (dashed line) and $\varepsilon = 0.4\omega_m$ (dot dashed line) by assuming $|\psi(0)>=|J,-J>$ ($J = 10$) and the oscillator to be initially prepared in a coherent state with an average of one quanta where $<x(0)>\neq 0$, $<p(0)>= 0$ and $\chi_x(0) = \chi_p(0) = 1$. (a): Plot of $\zeta_{S,x}(t)$ as a function of time. (b): Plot of $\zeta_{S,y}(t)$ as a function of time. The other parameters used are $g_0 = 0.8\omega_m$, $\gamma = 0$ and $\beta = 10$.

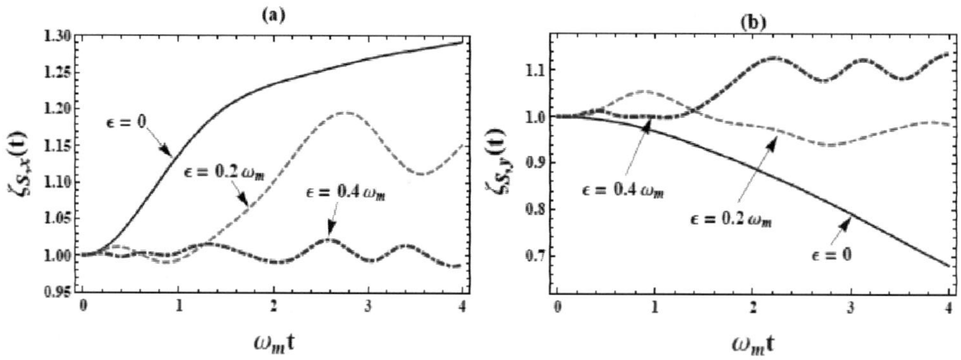

Figure 3. (color online) Squeezing parameter $\zeta_S(t)$ versus scaled time ($\omega_m t$) in absence of cavity-mirror coupling $\varepsilon = 0$ (solid line) and for two different values of mirror-photon coupling with $\varepsilon = 0.2\omega_m$ (dashed line) and $\varepsilon = 0.4\omega_m$ (dot dashed line). We assume $|\psi(0)>=|J,-J>$ ($J = 10$) and the oscillator is initially prepared in an amplitude squeezed vacuum state where $<x(0)>=<p(0)>= 0$ and $\chi_x(0) = 0.3807$. (a): Plot of $\zeta_{S,x}(t)$ as a function of time. (b): Plot of $\zeta_{S,y}(t)$ as a function of time. The other parameters used are $g_0 = 0.8\omega_m$, $\gamma = 0$ and $\beta = 10$.

$$\zeta_{S,x}(t)=\sqrt{\frac{2}{J}}\Delta J_x, \tag{4}$$

$$\zeta_{S,y}(t)=\sqrt{\frac{2}{J}}\Delta J_y. \tag{5}$$

The state vector for the combined spin ensemble held within the optomechanical cavity and the single-mode quantized cavity field can be given as:

$$|\Psi_1(t)>=\sum_{n,M}C_{n,M}(t)|n>|M>, \tag{6}$$

where we use $|M>$ for $|J,M>$ since J is a constant of motion for our Hamiltonian and $C_{n,M}(t)$ represent the time-dependent coefficients. The harmonic-oscillator eigen states are represented by $|n>$ with $a|n>=\sqrt{n}|n-1>$ and $a^\dagger|n>=\sqrt{n+1}|n+1>$. Furthermore, the harmonic oscillator and BEC are assumed to be uncorrelated for the initial state of the system such that the state vector can be written as a direct product

$$|\Psi_1(0)>=\left(\sum_n C_n(0)|n>\right)|\psi(0)>, \tag{7}$$

where, $|\psi(0)>=|J=N/2,M=-N/2>$. The initial harmonic oscillator number state coefficients are given by $C_n(0)$ [29]. The basis of harmonic-oscillator states are also to be truncated of course by determining an adequate number of states until the result is not changing. The equation of motion for Hamiltonian (3) is evaluated using Schrodinger's equation and is given as:

$$\begin{aligned}i\dot{C}_{n,M}(t)=&\left[\frac{\gamma}{N}M^2+g(t)n\right]C_{n,M}(t)+\frac{g_0}{2\sqrt{N}}\left[\sqrt{n}\sqrt{J(J+1)-M(M+1)}\right]C_{n-1,M+1}(t)\\&+\frac{g_0}{2\sqrt{N}}\left[\sqrt{n+1}\sqrt{J(J+1)-M(M-1)}\right]C_{n+1,M-1}(t)\end{aligned}. \tag{8}$$

The value of J is considered to be 10. Interaction Hamiltonian couples only those states where $n+M$ is always a constant. The time-dependent state vector for the system can be written as a sum over all the possible eigen states. In order to obtain $|\Psi_1(t)>$, time-dependent coefficients $C_{n,M}(t)$ are evaluated using equation (8) with the help of Mathematica 9.0. Thus, squeezing parameters defined in equations (4) and (5) can be evaluated with time using this state vector of the system. Experimentally, the frequency of mechanical oscillator may vary from 2π × 100Hz [30], 2π × 10kHz [31] to 2π × 73.5MHz [32]. In an optomechanical system, the mirror-photon coupling rate can take the values as low as 2π×2.7Hz [33] to as high as 2π × 4 × 10^5Hz [34]. The intracavity field interacting with the cloud of BEC held within the optomechanical cavity may have a coherent coupling strength of 2π × 5.86kHz [35] (2π × 10.9MHz [36]). Moreover, in our calculations, if the frequency of mechanical resonator is varied from 2π × 100Hz to 2π × 10kHz, then, the time will come out to be nearly of the order of milliseconds. There have been many experimental and theoretical papers in which the time evolution of spin-squeezing parameter is usually taken to be small (nearly of the order of milliseconds) [3, 19, 37–39]. At short times, it should be possible to avoid the effects of damping and thermal noise such that the substantial amount of spin-squeezing can be achieved at an earlier time.

In figure 2, we plot $\zeta_S(t)$ versus scaled time ($\omega_m t$) in the absence of cavity-mirror coupling $\varepsilon=0$ (solid line) and for two different values of mirror-photon coupling ε, $\varepsilon=0.2\omega_m$ (dashed line) and $\varepsilon=0.4\omega_m$ (dot dashed line). Here, we assume the harmonic oscillator to be initially prepared in a coherent state. Figure 2(a) shows that even though we start with a coherent state, squeezing along the x-direction increases significantly with the increase in light-mirror coupling. This is due to the fact that the radiation pressure force acting on the end mirror as a result of coupling between the optical field and the mirror motion generates squeezing in harmonic oscillator, known as pondermotive squeezing [14], which is then transferred to the condensate atoms via interaction Hamiltonian. However, spin-squeezing along the y-direction deteriorates with the increase in mirror-photon coupling and shows the maximum amount of squeezing for $\varepsilon = 0$ (see figure 2(b)). Thus, squeezing in spins along both the x- and y- directions can be significantly altered by choosing the appropriate value of optomechanical coupling. Further, we observe that spins in the x-direction remains squeezed for sufficiently large times (persistent spin-squeezing), with relatively much higher degree of persistent spin-squeezing for $\varepsilon = 0.4\omega_m$.

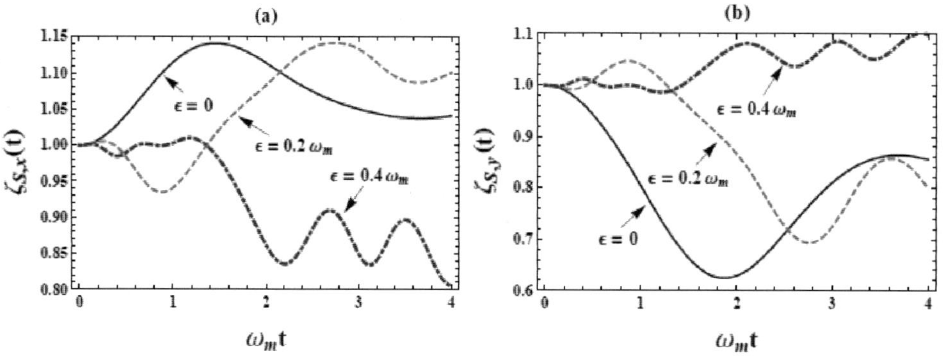

Figure 4. (color online) Squeezing parameter $\zeta_S(t)$ versus scaled time ($\omega_m t$) in absence of cavity-mirror coupling $\varepsilon = 0$ (solid line) and for two different values of mirror-photon coupling with $\varepsilon = 0.2\omega_m$ (dashed line) and $\varepsilon = 0.4\omega_m$ (dot dashed line) by assuming $|\psi(0)\rangle = |J,-J\rangle$ ($J = 10$) and the oscillator to be initially prepared in an amplitude squeezed state where $< x(0) > \neq 0$, $< p(0) > = 0$ and $\chi_x(0) = 0.3904$. (a): Plot of $\zeta_{S,x}(t)$ as a function of time. (b): Plot of $\zeta_{S,y}(t)$ as a function of time. The other parameters used are $g_0 = 0.8\omega_m$, $\gamma = 0$ and $\beta = 10$.

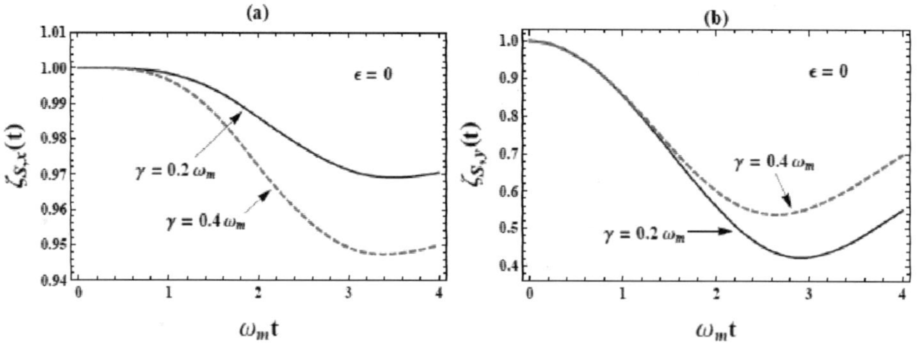

Figure 5. (color online) Squeezing parameter $\zeta_S(t)$ versus scaled time ($\omega_m t$) in the absence of optomechanical coupling $\varepsilon = 0$ for two different values of nonlinear atomic interaction strength with $\gamma = 0.2\omega_m$ (solid line) and $\gamma = 0.4\omega_m$ (dashed line) by assuming $|\psi(0)\rangle = |J,-J\rangle$ ($J = 10$) and the oscillator to be initially prepared in a coherent state with an average of one quanta where $< x(0) > \neq 0$, $< p(0) > = 0$ and $\chi_x(0) = \chi_p(o) = 1$. (a): Plot of $\zeta_{S,x}(t)$ as a function of time. (b): Plot of $\zeta_{S,y}(t)$ as a function of time. The other parameters used are $g_0 = 0.8\omega_m$ and $\beta = 10$.

Figure 3 illustrates the time evolution of spin-squeezing parameter for the oscillator to be initially prepared in an amplitude squeezed vacuum state. As before, we observe that higher value of mirror-photon coupling favors spin-squeezing in the x-direction while lower value of ε favors spin-squeezing along the y-direction. However, no squeezing along x-direction is observed in the absence of cavity-mirror coupling. Further note that the persistent squeezing along x-direction is observed for high ε, whereas, persistent and high squeezing is attained in the y-direction for $\varepsilon = 0$. We further show the variation of $\zeta_S(t)$ with time in figure 4 by considering the oscillator to be initially prepared in an amplitude squeezed state. These figures illustrate a similar kind of behavior as in the previous cases but more important is the fact that, in the presence of optomechanical coupling, initially squeezed state is able to produce a higher degree of persistent spin-squeezing in the x-direction as compared to previous cases. This is due to the pondermotive force acting on the movable mirror (cavity boundary) that helps in producing further squeezing in cavity mode, which gets drained away and then transferred to the spins. However, maximum squeezing along y-direction is attained in the absence of optomechanical coupling for the initial coherent state. This implies that the atomic spin-squeezing along both the x- and y- directions can be coherently controlled by the mirror-photon coupling. In principle, it is evident that spin-squeezing in two-level atoms can also be generated from nonlinear atomic interaction [6, 18, 19]. Thus, keeping other parameters same as before, we have also examined the effect of nonlinear atom-atom interaction γ on spin-squeezing in the absence of optomechanical coupling $\varepsilon = 0$ for the initial coherent cavity field (figure 5). We have observed a respective decrease and increase in the value of squeezing parameter $\zeta_{S,x}(t)$ and $\zeta_{S,y}(t)$ with increase in γ. In a realistic optomechanical system, decay of the cavity mode and thermal noise acting on the mechanical resonator will deteriorate the pondermotive squeezing and consequently the atomic spin-squeezing. However, significant squeezing can still be obtained if pondermotive interaction predominates over thermal noise [40], which can be achieved by using a large enough cavity finesse and mechanical quality factor such that substantial squeezing can still be achieved during the initial dynamics.

4. Conclusion

We have shown the effect of mirror-photon coupling on the generation of squeezed-spin states for a BEC confined in a lossless optomechanical cavity. In particular, we have studied the squeezing in spins for the three special cases- initial cavity field in coherent state, squeezed vacuum state and squeezed state. We have observed that the spin-squeezing in both the x- and y- directions can be coherently controlled by choosing the appropriate value of optomechanical coupling. Persistent spin-squeezing can also be generated by the proper choice of the system parameters. The pondermotive force acting on the movable mirror as a result of coupling between the optical field and the movable mirror generates squeezing in the optical cavity field which gets transferred to the condensate atoms due to a Tavis-Cummings type interaction and results in squeezed-spin states. We found that, even if the initial state of the cavity field is coherent, still a high degree of squeezing is generated and transferred to the atoms, with maximum squeezing along the y-direction in the absence of optomechanical coupling. Maximum amount of spin-squeezing along the x-direction is observed in the presence of optomechanical coupling for the initial cavity field prepared in the amplitude squeezed state. Such correlated-particle states have applications in entanglement detection as well as high precision metrology.

References

[1] Kitagawa M and Ueda M 1993 *Phys. Rev. A* **47** 5138
[2] Wineland D J, Bollinger J J, Itano W M, Moore F L and Heinzen D J 1992 *Phys. Rev. A* **46** R6797
[3] Wineland D J, Bollinger J J, Itano W M and Heinzen D J 1994 *Phys. Rev. A* **50** 67
[4] Polzik S 2008 *Nature* **453** 45

[5] Cronin A D, Schmiedmayer J and Pritchard D E 2009 *Rev. Modern Phys.* **81** 1051
[6] Sørensen A, Duan L M, Cirac J I and Zoller P 2001 *Nature* **409** 63
[7] Bigelow N 2001 *Nature* **409** 27
[8] Guehne O and Toth G 2009 *Phys. Rep.* **474** 1
[9] Turchette Q A, Wood C S, King B E, Myatt C J, Leibfried D, Itano W M, Monroe C and Wineland D J 1998 *Phys. Rev. Lett.* **81** 3631
[10] Wang X, Sanders B C 2003 *Phys. Rev. A* **68** 012101
[11] Kimble H J and Walls D F 1987 Special issue on squeezed states of electromagnetic field *J. Opt. Soc. Am. B* **4** 1450
[12] Loudon R and Knight P L 1987 Special issue on squeezed light *J. Mod. opt.* **34** 709
[13] Kimble H J 1991 in Fundamental systems in Quantum Optics ed. Dalibard J, Raimond J M and Zinn-Justin J (Elsevier, New York)
[14] Corbitt T, Chen Y, Khalili F, Ottaway K, Vyatchanin S, Whitcomb S and Mavalvala N 2006 *Phys. Rev. A* **73** 023801
[15] Purdy T P, Yu P L, Peterson R W, Kampel N S and Regal C A 2013 *Phys. Rev. X* **3** 031012
[16] Brooks D W C, Botter T, Schreppler S, Purdy T P, Brahms N and Stamper-Kurn D M 2012 *Nature* **488** 476
[17] Julsgaard B, Kozhekin A and Polzik E 2001 *Nature* **413** 400
[18] Orzel C, Tuchman A K, Fenselau M L, Yasuda M and Kasevich M A 2001 *Science* **291** 2386
[19] Riedel M F 2010 *Nature* **464** 1170
[20] Estève J, Gross C, Weller A, Giovanazzi S and Oberthaler M K 2008 *Nature* **455** 1216
[21] Jaaskelainen M, Meystre P 2006 *Phys. Rev. A* **73** 013602
[22] Pinard M, Hadjar Y and Heidmann A 1999 *Eur. Phys. J. D.* **7** 107
[23] Bhattacherjee A 2009 *Phys. Rev. A* **80** 043607
[24] Li S C, Fu L B and Liu J 2011 *Phys. Rev. A* **84** 053610
[25] Sakurai J J 1993 *Modern Quantum Mechanics* (New York: Addison-Wesley)
[26] Arecchi F T, Courtens E and Gilmore R 1972 *Phys. Rev. A* **6** 2211
[27] Scully M O and Zubairy M S 1997 *Quantum Optics* (Cambridge University Press)
[28] Mahajan S, Aggarwal N, Bhattacherjee A B and Mohan M 2014 *Chin. Phys. B* **23** 020315
[29] Henry R W and Glotzer S C 1988 *Am. J. Phys.* **56** 318
[30] Cole G D 2012 *Proc. SPIE* **8458** 845807
[31] Hunger D, Camerer S, Hänsch T W, König D, Kotthaus J P, Reichel J and Treutlein P 2010 *Phys. Rev. Lett.* **104** 143002
[32] Schliesser A, Rivière R, Anetsberger G, Arcizet O and Kippenberg T J 2008 *Nature Phys.* **4** 415
[33] Gigan S, Böhm H R, Paternostro M, Blaser F, Langer G, Hertzberg J B, Schwab K C, Bäuerle D, Aspelmeyer M and Zeilinger A 2006 *Nature* **444** 67
[34] Chan J, Eichenfield M, Camacho R and Painter O 2009 *Opt. Express* **17** 3802
[35] Nagorny B, Elsässer T, Richter H, Hemmerich A, Kruse D, Zimmermann C and Courteille P 2003 *Phys. Rev. A* **67** 031401
[36] Brennecke F, Ritter S, Donner T and Esslinger T 2008 *Science* **322** 235
[37] Li Y, Treutlein P, Reichel J and Sinatra A 2009 *Eur. Phys. J. B* **68** 365
[38] Helmerson K and You L 2001 *Phys. Rev. Lett.* **87** 170402
[39] Sintara A, Dornstetter J C and Castin Y 2012 *Front. Phys.* **7** 86
[40] Vitali D and Tombesi P 2011 *C. R. Physique* **12** 848

Journal of Physics: Conference Series **672** (2016) 012007 doi:10.1088/1742-6596/672/1/012007

Detection of CO_2 laser radiation in a ferrite

H Haroyan[1], A Makaryan[1], K Movsisyan[1], Farokh Nazari[1], V Tadevosyan[1], H Julfayan[2]

[1] Microwave Radiophysics and Telecommunications Dept., Yerevan State University, Alex Manoogian 1, 0025, Yerevan, Armenia

[2] Institute of Radiophysics and Electronics, Alikhanian Brothers str. 1, 0203 Ashtarak, Armenia

E-mail: armenm@ysu.am

Abstract. The detection of plane polarized, amplitude modulated radiation of CO_2 laser ($\lambda \sim 10$ μm) in a transparent ferrite at room temperature is experimentally performed. It is shown that the detected signal is observed when the magnetization of the ferrite reaches the saturation value. The dependence of detected signal parameters on the external magnetic field is well correlated with the magnetization curve of the ferrite sample.

Ferromagnetism in general is widely used in modern industry and technology, underlying as a basis for many electrical and electromechanical devices such as electromagnets, electric motors, generators, transformers, and magnetic storage such as tape recorders, and hard disks. Ferromagnetic materials are widely used in electronics due to wide range of their physical properties. In addition to their usage in customer low-frequency electronics, ferrites found numerous applications in microwave and optoelectronic devices. In many studies the possibility of generation, detection, frequency conversion, and amplification of microwave radiation has been demonstrated [1].

The nonlinear behavior of the magnetic susceptibility of ferromagnets exposed to radiofrequency and microwave fields has been well investigated [1], [2]. These phenomena have a great scientific and practical value in the areas of signal detection, frequency conversion, etc.

Nevertheless, despite the fact that there is a plurality of ferrites, which are transparent in the infrared and optical domain, and are successfully used for rotation of polarization plane, for radiation control, etc., the nonlinear features of ferromagnets in this range are not well investigated. There are a series of works devoted to reorientation of the magnetic momentum in ferromagnetic materials exposed to ultrashort laser pulses (see e.g. [3]). However, it is generally considered that ferromagnetic materials cannot exhibit magnetic nonlinearity, because the magnetic permeability in said bands is equal to unity. According to [3] the ultrafast variation of the magneto-optical response taking place during an ultrashort laser pulse can be explained in terms of nonlinear optical susceptibility. However, the appearance of nonlinear optical susceptibility of a ferromagnet in an external magnetic field is not physically substantiated. This description is purely phenomenological, and interpretation of ultrafast magneto-optical response of ferromagnetic materials is still a subject for debates.

In [4] the optical detection of femtosecond laser radiation (wavelength range 0.71-0.95 μm and 200 kW peak power) by the partially transparent YIG ferromagnet was obtained. Based on the fact that the dependence of detected signal parameters on external magnetic field is well correlated with

Journal of Physics: Conference Series **672** (2016) 012007 doi:10.1088/1742-6596/672/1/012007

the magnetization curve of the used ferromagnet sample, the authors suggested that the detection is attributed to nonlinear interaction of laser radiation with a ferromagnet. Meanwhile, if detection is feasibly due to nonlinearity of magnetisation curve, this phenomenon will be seen for any transparent ferromagnet medium, in the case of nonlinear operating range of magnetisation curve.

The equation of motion for the magnetic momentum M in an alternating magnetic field H was proposed in [4], where excitation of the magnetic momentum appears to be caused by the induction of magnetic field, also taking into account losses in the ferromagnet:

$$\frac{d\mathbf{M}}{dt} = -\gamma[\mathbf{M}\times\mathbf{H}] - \gamma^2 I \frac{d\mathbf{H}}{dt} + \boldsymbol{R}, \quad (1)$$

where $\gamma = q/2mc$ is the gyromagnetic ratio, I - moment of inertia, and R - dissipative term.

As it follows from equation (1), the alternating magnetic field of the wave can change the magnetic momentum of ferromagnet. Ferromagnetic materials exhibit magnetic momentum saturation with the increase of external magnetic field. As a result, clearly nonlinear response for magnetic momentum is expected when the applied magnetic field changes in the proximity of the saturation level. Thus, in the case of penetration of the electromagnetic wave in a ferromagnetic material, in certain conditions one may expect nonlinear interaction of the electromagnetic wave with the ferromagnet.

In this paper we report the results of experimental investigation of possibility to detect far infrared laser radiation in a transparent ferromagnet recorded due to the excitation of magnetic momentum of a magnetized ferromagnet by induction of magnetic field of the laser radiation.

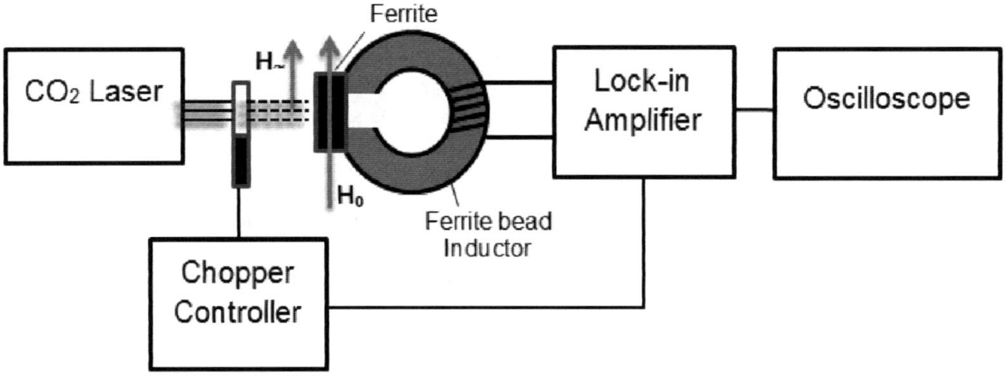

Figure 1. Schematic diagram of the experimental setup. H_0 shows direction of the external magnetic field, and $H_{\sim}$ denotes the magnetic field of laser radiation.

Schematic diagram of the experimental setup is sketched in Figure 1. The continuous-wave linearly polarized radiation of CO_2 laser ($\lambda \sim 10$ μm) with output power of about 15 W was used as an electromagnetic field. The laser radiation was modulated by a mechanical chopper in order to realize lock-in detection of the recorded signal. The sample of NM2000 ferrite widely applicable in the low-frequency electronics served as a ferromagnet. This sample was partially transparent in far infrared: the measured absorption coefficient was 16 cm^{-1} for $\lambda \sim 10$ μm wavelength. To record variation of the magnetic momentum in the sample, the latter was attached to a ferrite bead inductor with $n = 500$ turns winding (see Figure 1). The output signal of the inductor was detected by a lock-in amplifier and recorded by an oscilloscope.

In Figure 2 the magnetization curve of NM2000 ferrite and the dependence of the detected signal on magnetic field are plotted. In Figure 2b for evident demonstration of the dependence of detected signal on the non-linearity of the magnetization curve the second order derivative of the magnetization curve is shown.

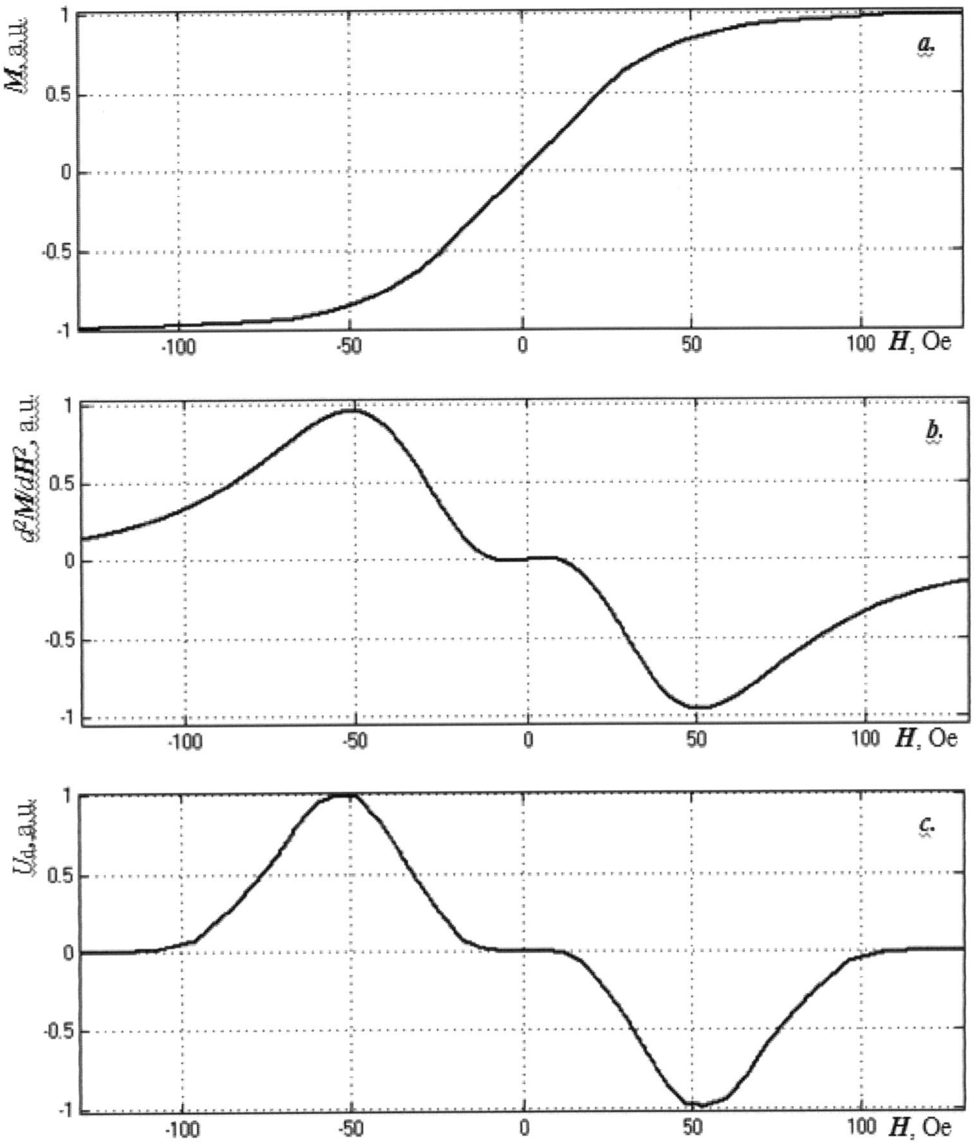

Figure 2. ***a.*** Magnetization curve; ***b.*** second order derivative of the magnetization curve, ***c.*** magnitude of the detected signal versus external magnetic field.

As is seen from Figure 2, no signal detected in the absence of an external magnetic field. Signal reaches its maximum value in the region where the change of the slope of magnetization curve for a ferrite sample is maximal. Moreover, the reversal of external magnetic field alters the polarity of the detected signal. $U_{max} \approx 150$ μV peak amplitude of the detected signal was recorded at $H_0 \approx 50$ Oe.

It should be noted that besides direct detection of far infrared radiation, nonlinear interaction of laser radiation with ferromagnets can find practical application in frequency conversion of laser radiation, recording and storage of information, etc.

References

[1] Gurevich A G 1961 *Ferrites in nonlinear microwave devices, collection of articles* (Moscow: Foreign Literature Publishing House) in Russian
[2] Yakovlev Yu M and Gendelev S Sh 1975 *Sov. Radio* Moscow (in russian)
[3] Kimel A V, Kirilyuk A, Hansteen F, Pisarev R V and Rasing Th 2007 *J. Phys.: Condens. Matter* **19** 04320
[4] Martirosian R M *et al* 2014 *JETP Letters* **99** 8 435

Free carrier absorption in quantum well in consideration of the interaction with impurities

A H Gevorgyan, E M Kazaryan and A A Kostanyan

Institute of Mathematics and High Technologies, Russian-Armenian University, Hovsep Emin Street, 0051 Yerevan, Armenia

Abstract. Intrasubband transitions caused by light absorption in a parabolic quantum well is considered taking into account the scattering by ionized impurity centers. To calculate the scattering matrix element Born approximation is used, and the interaction with the impurity is discribed via screened Coulomb potential. An analytical expression for the absorption coefficient is obtained based on the the initial absorption of a photon and a further scattering on ionized impurity center. Absorption coefficient frequency characteristics and dependence on the width of the quantum well is examined.

1. Introduction

It is known that the direct light absorption by free carriers is impossible, since it contradicts conservation laws of energy and momentum. Availability of phonons, impurities and other lattice defects enable light absorption, since scattering on a third particle provides a necessary change in a momentum. Free carriers absorption (FCA) is an effective tool to identify and assess the mechanisms of scattering. FCA was reviewed in massive semiconductors in the second-order of perturbation theory, taking into account various mechanisms of scattering [1], including ionized impurities [2].

The most interesting case of FCA is in the low-dimensional structures. Owing to dimensional quantization (for example in one direction in quantum wells), energy subbands arise which make possible transitions not only in one subband (intrasubband transitions), but also transition between two subbands (intersubband transitions) [3].

Intrasubband transition in the quantum wells (QW) cause a great interest due to its unique characteristics: large dipole momentum, ultra-fast relaxation, great opportunity to control the transitions wave length [4-6]. It is important not only for the fundamental physics, but also for development of new technological applications. Many devices are developed based on intrasubband and intersubband transitions in heterostructures from QW. For example, infrared photodetectors [7, 8], superfast optical modulators [9], optical switches [10] and quantum cascade lasers [11].

One of the first theoretical works devoted to light absorption by free carriers in size quantized structures are the works of Kazaryan et al [12, 13], where within the second order of perturbation theory frequency dependency of absorption coefficient is obtained for nondegenerate electronic gas in QW and in quantum well wires. Scattering mechanisms on acoustic phonons [12] and the ionized impurity centers [13] are considered. Later FCA in QW had been studied by many authors, taking into account scattering on acoustic phonons [14], polar and non polar optical phonons [15,16,17], and also on the ionized screened impurities [18, 19]. In the studied works confinement potentials had been approximated by single rectangular, and also doubled [20] (applied to quantum cascade lasers) finite or infinite rectangular potentials.

On the other hand, there is a need to create more realistic model of the confinement potential which takes into account both physicochemical properties of structure and its geometry. The first form height and shape of the potential barrier on the interface, and the geometry defines symmetry of a Hamiltonian (in lack of external fields). For these purposes various models of the confinement potential for low-dimensional systems were applied [21-25]. As a first approximation the confinement potential can be approximated by parabolic one. With regard to the problem considered in the presented work it allows to obtain analytical expression for absorption coefficient. Let us notice, that ideal parabolic approximation of the confinement potential works fine for low energy levels. Further, it is assumed that confinement potential of QW looks like

$$V_{conf}\left(z\right)=\frac{m^{*}\omega_{0}^{2}z^{2}}{2}, \tag{1}$$

where m^{*} – effective mass of electron, ω_0 – frequency defined by virial theorem according to a relation $\omega_0 \sim \frac{\hbar}{m^{*}a^{2}}$, where a – is QW width. For the exact equality we will present ω_0 in a form

$$\omega_0 = \frac{\gamma\hbar}{m^{*}a^{2}}, \tag{2}$$

wher γ – some adjustable parameter providing exact equality in (2).

2. Theory

Calculations are carried out on the basis of the standard theory of quantum transitions, according to the general formula:

$$\alpha = \frac{\sqrt{\varepsilon}}{cN}\sum_{i} P_i f_i, \tag{3}$$

wher ε – dielectric constant, N – quantity of the photons falling on the QW in a unit of time per unit area, c – light speed, f_i – charge distribution function, P_i – transition probobility (count of transition per unit time).

Expression for P_i in the second order of the perturbation theory has an appearance [26-28]:

$$P_i = \frac{2\pi}{\hbar}\sum_{f}\left|\sum_{m}\frac{\left|M_{im}^{V}\right|\left|M_{mf}^{J}\right|}{E_i - E_m + \hbar\omega}\right|^{2}\delta(E_f - E_i - \hbar\omega), \tag{4}$$

where E_i – initial state energy, E_f – finit state energy, E_m – middle state energy, M_{im}^{V} – matrix element, caused by absorption of a photon, M_{mf}^{J} – matrix element, caused by scattering on ionized impurity centers. In z direction electron located in parabolic QW, and in the plane *(x, y)* there is a two-dimensional translational symmetry (see Figure.1).

It is supposed that incident light is linearly polarized, and incident angle is not specified. It should be mentioned, that for the intrasubband transitions one should demand that light incidence is not parallel to QW plane (vector of polarization is not perpendicular to QW plane), and vice cersa for intersubband transitions [3]. This requirement is equivalent to that if polarization vector *x, y* components are non-zero: $E_{x,y} \neq 0$ (figure 1).

In case of initial absorption of a photon (figure 2, transition *imf*), transition $i \to m$ is followed by absorption of a photon (Figure. 2), and the transition of $m \to f$ corresponds to the scattering on ionized impurity centers. At initial scattering (figure 2, transition *im'f*) we have: $i \to m'$ – scatteing on the impurity center and $m' \to f$ – photon absorbtion. m and m' states are virtual.

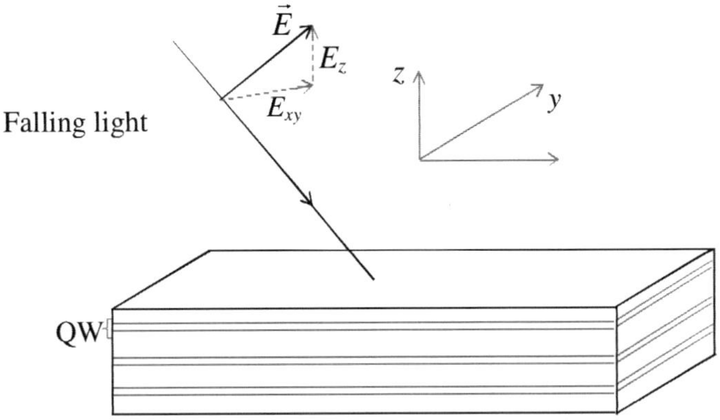

Figure 1. The schematic view of the falling on QW electromagnetic radiation.

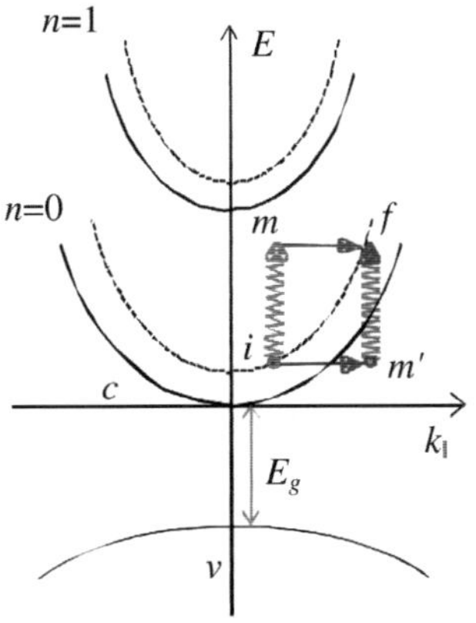

Figure 2. Diagram of energy transitionsn.

We use Born approximation for calculation of a scattering matrix element. We will consider the elastic mechanism of scattering on ionized impurity. In Figure 3 the change of a two-dimensional wave vector of an electron in case of elastic scattering mechanism is represented: $\vec{k}$ is a wave vector of an electron before interaction with impurity, and $\vec{k}\,'$ – after interaction. As it can be seen from figure the angle of a scattered wave vector lies in an interval $(-\pi, \pi)$. As it can be obtained from figure 3 the absolute value if of wave vectors difference can be presented as

$$\left|\vec{k}\,' - \vec{k}\right| = 2k \sin\frac{\theta}{2}, \quad (5)$$

θ – is scattering angle.

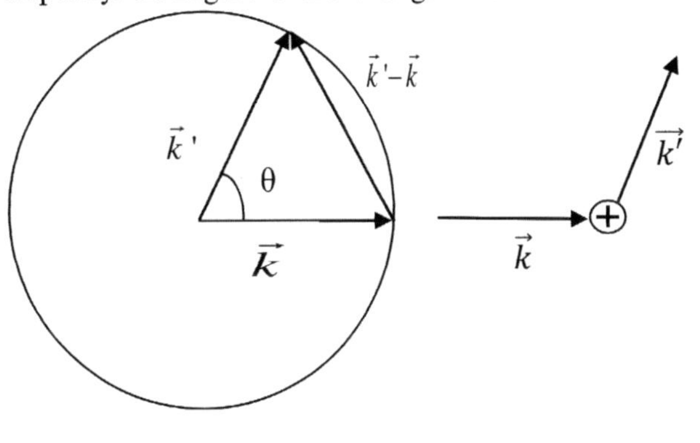

Figure 3. The elastic mechanism of scattering model

Wave function is

$$|\varphi\rangle = \sqrt{\frac{1}{S}}\frac{1}{\sqrt{2^n n!}}\left(\frac{\gamma}{\pi a^2}\right)^{1/4} e^{i\vec{k}_{//}\vec{\rho}} e^{-\frac{\gamma z^2}{2a^2}} H_n\left(\frac{\sqrt{\gamma}z}{a}\right), \quad (6)$$

where a – QW width, S – surface area, H_n – Hermite polynomials.

In further calculations we will consider that at first there is an absorption of a photon, and then scattering on the impurity center (pic. 2, transition of *imf*). The energy spectrum:

$$E_i = \frac{\hbar^2 k_i^2}{2m^*} + \frac{\hbar\omega_{osc}}{2}; \qquad E_m = \frac{\hbar^2 k_m^2}{2m^*} + \frac{\hbar\omega_{osc}}{2} + \hbar\omega;$$
$$E_f = \frac{\hbar^2 k_f^2}{2m^*} + \frac{\hbar\omega_{osc}}{2}, \qquad \hbar\omega_{osc} = \frac{\gamma\hbar^2}{m^* a^2} \quad (7)$$

where $\hbar\omega$ – photon energy. Note that processes like $i \to m \to f$ and $i \to m' \to f$ have equal probobilities.

Taking into account an expression (6) the matrix element caused by absorption of a photon has a form:

$$M_{im} = \langle \varphi_m | H' | \varphi_i \rangle = \frac{1}{aS}\sqrt{\frac{1}{\pi}} \int e^{-i\vec{k}_m\vec{\rho}} e^{-z^2/2a} H_m(z/a) \left[\frac{e}{mc}(\vec{A}\vec{p})\right] e^{i\vec{k}_i\vec{\rho}} e^{-z^2/2a} H_i(z/a) d\vec{r} =$$
$$= \frac{2\pi^2\sqrt{2\pi}}{S} \hat{e}\vec{k}_i \delta_{k_{//}^m k_{//}^i} \left(\frac{i\hbar e}{mc}\right) \frac{c}{\tilde{n}} \frac{\hbar^2}{\sqrt{\hbar\omega}}. \quad (8)$$

And the matrix element of scattering is:

$$M_{mf}^J = \langle \varphi_f | V_{Yukava} | \varphi_m \rangle = \int_0^\infty \int_0^{2\pi} \int_{-\infty}^{\infty} \varphi_f^* V_{Yukava} \varphi_m \rho d\rho d\varphi dz =$$
$$= \frac{2\pi}{S} \frac{Ze^2}{\varepsilon} \sqrt{\frac{1}{k^2 + \frac{1}{L_D^2}}} \exp\left(\frac{a^2}{4\gamma}\left(k^2 + \frac{1}{L_D^2}\right)\right), \quad (9)$$

where $k = |\vec{k}| = |\vec{k}_f - \vec{k}_m|$, L_D - is Debye screening length, and scattering potential is takne in the form (Debye-Hucke, Yukava)

$$V_{Yukava} = \frac{Ze^2}{\varepsilon\sqrt{\rho^2 + z^2}} \exp\left(-\frac{\sqrt{\rho^2 + z^2}}{L_D}\right); \quad \text{where} \quad L_D = \sqrt{\frac{\varepsilon k_B T a}{4\pi^2 e^2 n_e}}. \quad (10)$$

For absorption coefficient calculation (see eq. (4)) in case of initial absorption of a photon and further scattering on ionized impurity we use expression [29]

$$\alpha(\omega) = \frac{1}{N\cdot(c/\tilde{n})}\left(\frac{2S}{(2\pi)^2}\right)^2 \int d^2\vec{k} d^2\vec{k}' \left(\frac{2\pi}{\hbar} \frac{|M_{im}^V|^2 |M_{mf}^J|^2}{(E_i - E_m + \hbar\omega)^2} \delta(E_f - E_i - \hbar\omega)\right) f(k)[1 - f(k')], \quad (11)$$

where $f(k)$ and $f(k')$ are the probabilities of filling of initial and final states (Fermi-Dirac distribution), N – quantity of the incident photons per square (considered one-photon absorption), $\tilde{n}$ – refraction index.

We consider a nondegenerate electronic gas with a temperature T, neglecting $f(k')$ (i.e. $f(k') = 0$); $f(k)$ is the Boltzmann distribution function

$$f(k) = \frac{n_e}{N_c} e^{-\frac{E}{k_B T}} = \frac{n_e}{2L}\left(\frac{2\pi\hbar^2}{m^* k_B T}\right)^{3/2} e^{-\frac{E_i}{k_B T}}, \quad (12)$$

where n_e – concentration of 2D free electrons.

Substituting the expressions of the matrix elements (8), (10) in (11), and taking into account a distribution function (12), we will receive the following expression for absorption coefficient:

$$\alpha(\omega) = C\left(3 + \frac{\hbar\omega}{2E_T} + \frac{1}{16}\left(\frac{\hbar\omega}{E_T}\right)^2\right)\left(\frac{1}{\hbar\omega}\right)^{4.5} \exp\left(\hbar\omega\left(\frac{m^* a^2}{2\gamma\hbar^2} - \frac{1}{8E_T}\right)\right) \frac{1}{\left(1 + \frac{\hbar^2}{mL_D^2\hbar\omega}\right)}, \quad (13)$$

where we have used the following notations:

$$E_T = k_B T;$$
$$C = 1504 n_e N_I \frac{\hbar^2 e^6 E_T}{\varepsilon^{5/2} a m^{3/2} c} \exp\left(-\frac{\hbar\omega_{osc}}{2E_T} + \frac{a^2}{2\gamma L_D^2}\right), \quad (14)$$

N_I – concentration of the 2D ionized impurites, ε – the dielectric constant.

The difference between absorption coefficients between screened and pure Coulomb potentials can be presented as

$$\alpha_{Yukava}(\omega)=\alpha_{Coulomb}(\omega)\frac{\exp\left(\frac{a^2}{2\gamma L_D^2}\right)}{\left(1+\frac{\hbar^2}{mL_D^2\hbar\omega}\right)}. \tag{15}$$

3. Results

All numerical calculations are made for QW of *GaAs* with the following parameters: $m^*=0.067m_e$, $\varepsilon=13.8$, $E_R=5.275$ мэВ, $a_B=104$ Å. Are received the results for one type of processes – with initial absorption of a photon and further scattering on the ionized impurity. Since other possible sequence of transitions (with initial scattering on the ionized impurity) is equiprobable, it is natural to expect that the result should be the same. The analysis of frequency dependence of absorption coefficient shows dependence of $\alpha(\omega)\sim\frac{1}{\omega^{4.2}}$. Therefore, for parabolic QW the absorption coefficient with the increase of photon energy decreases faster in comparison with result for a massive sample $\sim\frac{1}{\omega^{3.5}}$ [1].

Absorption coefficient dependence on light frequency for various QW widths is presented on figure 4. It is visible, that the absorption shifts to the bigger values of photon energy with the reduction of QW width. Calculations are made for temperature *T*=77K, as at higher temperatures phonon scattering mechanism dominates.

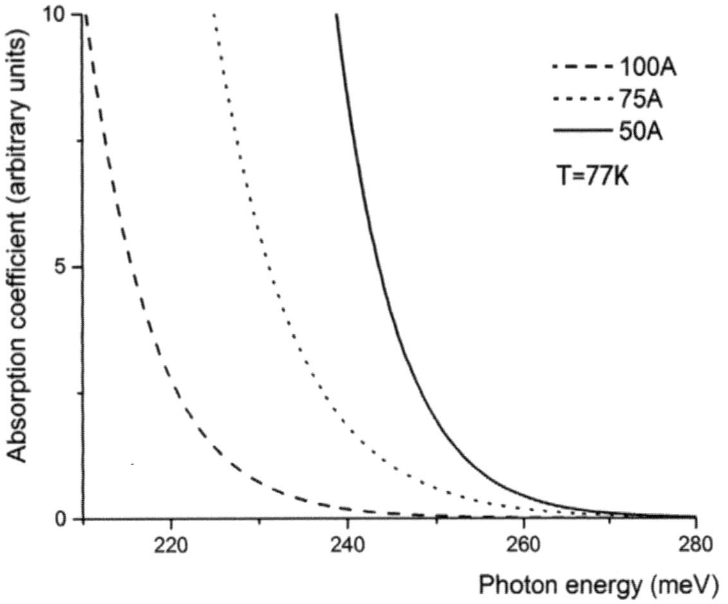

Figure 4. Dependence of absorption coefficient on incident photon energy at various values of QW widths.

References

[1] Fan H Y, Spitzer W and Collins R J 1956 *Phys. Rev.* **2** 101
[2] Rosenberg R and Lax M 1958 *Phys. Rev.* **12** 843

[3] Vorobyev *et al* 2000 *Optical phenomena in low dimensional semiconductor structures* (Saint-Petersberg: SPbSTU)
[4] Asano T, Noda S and Sasaki A 1998 *Physica E* **2** 111-115
[5] Paiella R 2006 *Intersubband Transitions in Quantum Structures* (McGraw-Hill Companies) pp. 137-140
[6] Helm M 2000 *Semiconductors and Semimetals* **62** 1-32, 73-80
[7] Alves F D P, Karunasiri G, Hanson N, Byloos M, Liu H C, Bezinger A and Buchanan M 2007 *Infrared Phys. & Technol.* **50** 182-186
[8] Li S S 2002 *Int. J. of High Speed Electronics and Systems* **12** 761-801
[9] Carter S G, Ciulin V, Sherwin M S, Hanson M, Huntington A, Coldren L A and Gossard A C 2004 *Appl. Phys. Lett.* **84** 840–842
[10] Iizuka N, Kaneko K and Suzuki N 2006 *IEEE J. Quantum Electr.* **42** 765-771
[11] Chakraborty T and Apalkov V M 2003 *Adv. Phys.* **52** 455-521
[12] Kazaryan E M, Grigoryan V G and Kazaryan A M 1976 *Fizika AS Arm. SSR* **11** 351-359
[13] Kazaryan E M and Aramyan K S 1976 *Fizika AS Arm. SSR* **11** 122-127
[14] Lee J and Spector H N 1983 *J. Appl. Phys.* **54** (7) 3921
[15] Adamska H and Spector H N 1984 *J. Appl. Phys.* **56** 1123-1127
[16] Kubakaddi S S and Mulimani B G 1986 *J. Phys.* **19** 11300-11306
[17] Bhat J S, Kubakaddi S S and Mulimani B G 1992 *J. Appl. Phys.* **72** 4966
[18] Sankeshwar N S, Kubakaddi S S and Mulimani B G 1989 *J. Phys* **32** 149-159
[19] Gashimzade F M 1990 *Phys Stat. Sol.b* **160** 177
[20] Carosella F, Ndebeka-Bandou C, Ferreira R, Dupont E, Unterrainer K, Strasser G, Wacker A and Bastard G 2012 *Phys. Rev. B* **85** 085310
[21] Petrosyan L S 2002 *Journal of Contemporary Physics (Armenian Academy of Sciences)* **37** 173-177
[22] Hayrapetyan D B, Kazaryan E M, Petrosyan L S and Sarkisyan H A 2015 *Physica E* **66** 7-12
[23] Hayrapetyan D B, Kazaryan E M, Kotanjyan T V and Tevosyan H K 2015 *Superlattices and Microstructures* **78** 40-49
[24] Hayrapetyan D B, Kazaryan E M and Tevosyan H K 2013 *Superlattices and Microstructures* **64** 204-212
[25] Maksym P A and Chakraborty T 1990 *Phys. Rev. Lett.* **65** 108
[26] Bastard G 1989 *Wave mechanics applied to semiconductor heterostructures* (Les editions de Physique)
[27] Prudnikov A P, Brychkov Yu A and Marichev O I 1986 *Integrals and Series* vol. **3** (Moscow: Nauka)
[28] Anselm A and Samokhvalov M 1978 *Introduction to Semiconductor Theory* (New York: Prentice Hall)
[29] Seeger K 1973 *Semiconductor Physics* (Wien: Springer-Verlag)

Dielectric confinement influenced screened Coulomb potential for a semiconductor quantum wire

K H Aharonyan[1,2] and N B Margaryan[1]

[1] National Polytechnic University of Armenia
[2] Armenian State Pedagogical University after Khachatur Abovyan

E-mail: ahkamo@yahoo.com

Abstract. A formalism of the Thomas-Fermi method has been applied for studying the screening effect due to quasi-one-dimensional electron gas in a semiconductor cylindrical quantum wire embedded in the barrier environment. With taking into account of strongly low dielectric properties of the barrier material, an applicability of the quantum wire effective interaction potential of the confined charge carriers has been revealed. Both screened quasi-one-dimensional interaction potential and effective screening length analytical expressions are derived in the first time. It is shown that in the long wavelength moderate limit dielectric confinement effect enhances strength of the screening potential depending on the both radius of the wire and effective screening length, whereas in the long wavelength strong limit the screening potential solely is determined by barrier environment dielectric properties.

1. Introduction

Wide interest to electron gas (EG) properties in quasi-one-dimensional (Q1D) semiconductor structures (quantum wires (QWR), nanorods, nano-whiskers or carbon nanotubes [1-4]) is initiated by application potential of these structures in high-speed electronic and optical devices [5]. A Q1D EG in these systems can be produced by intense optical illumination of a undoped QWR's or by modulating of the doping material [6]. An advanced selective doping makes it possible to spatially separate free carriers from the parent ionized impurities and to investigate charged Q1D carrier system rather than the neutral electron-hole plasma. As follows, a Q1D electron channel, confined in Q1D nanostructure (quantum confinement (QC)), can coexist with the confined coulomb centers (photo excited excitons, impurities) as in Q2D case [7] and screen Coulomb polarization fields between quasiparticles. In this sense, both the depth and wide scope of physics of the low-dimensional systems is mainly caused by strong quasiparticle correlations and weak screening of Coulomb interaction. Meanwhile the latter's strength and further enhancement are connected in the further lowering of the dimensionality of the nanostructure. In the 3D system Coulomb screened potential decays exponentially in real space and effectively becomes a short-range potential [8]. In a pure 2D structures Coulomb statically screened potential at the long-wavelength limit possess the 2D Coulomb law for the intermediate distances and power law spatial dependence for large distances, respectively [9, 10]. There is specific screening radius here and, in addition, screening parameter saturates both for the small temperature and large planar carrier density of the 2D EG. The Coulomb screening effect reducing takes place because free carriers can only screen Coulomb field between any two charges inside a low-dimensional system, whereas a part of field extending outside the confinement region remains unscreened. It is assumed that this tendency should go on more strong by for the transition from quantum well (QW) to QWR.

The physical concept, that dielectric screening would be insufficient to reduce Coulomb repulsion between 1D (Q1D) confined charge carriers originates from the Kuper's work [11] and later this model reinvestigated in Ref.[12] by Davis. As a main result, the screened Coulomb potential expressions inside and outside the ***conducting*** 1D single-wire and coaxial cylinder are obtained in the Tomas-Fermi approximation (TFA). Subsequently, Lee and Spector [13] used random-phase approximation (RPA) to evaluate a screening potential of the charged impurity in ***semiconducting*** wire. Reyes et al. [14] derived both the dielectric function of 1D EG and asymptotic forms of the pure 1D screened Coulomb potential in semiconductor QWR by using 1D generalization of the TFA. Recently, Wang et al. [15] derived an analytical formula for the screened Q1D Coulomb potential in semiconductor Q1D nanostructures.

A realistic semiconductor QWR either can be freestanding or surrounded with a barrier environment. In this sense, for the Coulomb center-based structures, use of the low dielectric constant barrier environment of semiconductor QWR (lower than the semiconductor dielectric constant) is favorable, as it enhances the Coulomb interaction inside the QWR (dielectric confinement effect (DC)) [16-18]. The latter enhances both electronic and optical properties of the nanostructure and as a major result it becomes possible to implement "Coulomb interaction engineering" therein [16].

This tunability of the Coulomb interaction in the QWR due to the barrier environment dielectric properties enhances and modifies many body effects as well, in particular, dielectric screening effect of the Q1D EG. Most of the theoretical works in this area have been done by RPA framework to calculate the linear response of the QWR to an external charge in presence of the DC effect. Wendler and Grigoryan [19] by using a two-subband model an influence of the DC effect on the Q1D intra- and intersubband plasmon have studied. It has been shown that for a 1D long wave limit intrasubband plasmon possess a linear dispersion, is independent of the dielectric screening of the QWR and is wholly screened by surrounding barrier environment as in the Q2D intrasubband case [20]. Whereas by Aharonyan et al. in presence of the strong DC effect has been shown [21], that together with the previous result, which is consistent for the case of very small wave vectors only, there exists 1D moderate small wave vectors range in long wave limit as well, that collective plasmon frequencies are independent of the 1D wave vector and increase with decreasing of the QWR radius. Recently in Ref.[22] it is found that in the long-wavelength limit the dielectric screening and collective excitations of the Q1D EG are to be strongly influenced by barrier environment and exclusion of the DC effect in free-carrier screening results in an erroneous charged impurity scattering rate in QWR.

In the present paper we explore a general problem of determining the interaction potential of point charges pair confined within the semiconductor QWR embedded in the dielectric barrier environment and screened by Q1D EG. Our treatment sharply differs from that derived in Refs.13-15, where assumes that Q1D EG has homogeneous background having dielectric constant same as the barrier media.

2. CQWR model and TFA 3D screening potential

We are considering an infinitely long semiconductor cylindrical QWR (CQWR) of a radius R filled by the active material with the dielectric constant ε_w and immersed in a dielectric barrier environment with the dielectric constant ε_b . We have to argue, that the QWR infinite length concept allows us to discuss definite realistic semiconductor narrow samples, where one can used an adiabatic approximation of the Coulomb interaction. The latter replaces the three-dimensional potential of electrons and holes by a 1D Coulomb potential that describes their interaction along the QWR axis. In this sense, we are omitting a discussion in certain semiconductor quantum rod systems, where both the transverse and longitudinal dimensions have comparable sizes (for more details see for example [23]).

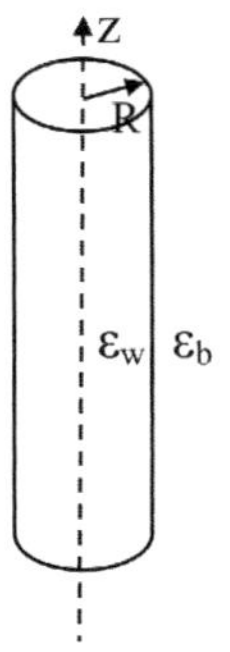

Figure 1. An infinitely long semiconductor cylindrical QWR (CQWR) of a radius R filled by the active material with the dielectric constant ε_w and immersed in a dielectric barrier environment with the dielectric constant ε_b.

Let us take in the cylindrical polar coordinates (ρ,φ,z), where the z axis coincides with the CQWR axis. The sought interaction screened potential $V_S(\rho, z) = e\Phi_S(\rho, z)$ of charges $-e$ and e to be located at points (0, 0) and (ρ, z) respectively, is related to the induced charge density by generalized Poisson's equation

$$\begin{cases} \nabla^2\Phi_S(r) = -\dfrac{4\pi e}{\varepsilon_w}\left[\delta(r) - \Delta n^{ind}(r)\right], & \rho \le R \\ \nabla^2\Phi_S(r) = 0, & \rho > R \end{cases} , \tag{1}$$

where $\Delta n^{ind}(\rho, z) = n^{ind}[\mu_0 + e\Phi_S(\rho, z)] - n^{ind}(\mu_0)$ is the variation of the density of EG, μ_0 is the chemical potential in the absence of the external perturbing field. Due to the cylindrical and reflection symmetry of the discussed problem Eq.(1) becomes independent of the angular coordinate φ.

In the equation system (1) we impose a constraint on the free electrons: the uniform local number density of electrons has to be taken zero everywhere except in the interior of a CQWR. It is supposed as well that only a charged channel contributes to the CQWR screening in presented model.

In the assumptions the externally applied charge $-e$ produces a linear response in Q1D EG, for large enough distances when $e\Phi_S << \mu_0$ we may expand $\Delta n^{ind}(\rho, z)$ and obtain in leading order as

$$\Delta n_V(\rho, z) = \frac{\partial n^{ind}}{\partial \mu_0} e\Phi_S(\rho, z) \cdot \tag{2}$$

Together with the abovementioned assumptions a strong confinement condition would be assumed that the quantum wire radius is small compared with the Coulomb center Bohr radius $a_0 = \varepsilon_w \hbar^2 / m_e e^2$ for bulk samples ($a_0 >> R$). As follows, the distances along the wire axis $|z| >> R$ are essential in discussed case and therefore the long wave 1D region $q << R^{-1}$ could be appropriate for the interaction screened potential in accordance with the TFA. This permits us to apply a geometrical adiabatic method introduced for the first time in Ref.[9] when discussing the Q2D EG properties of the QW system. Substituting equation (2) in the first equation of system (1) and expressing the screening potential $V_S(\rho, z)$ in Fourier components as $V_S(\rho,z) = \int_0^\infty V_S(\rho,q) e^{iqz}\, dq$ we will have

$$\left(\frac{\partial^2}{\partial\rho^2} + \frac{1}{\rho}\frac{\partial}{\partial\rho}\right) V_S(\rho,q) - \left[q^2 + \frac{4\pi e^2}{\varepsilon_w}\frac{\partial n^{ind}}{\partial\mu_0}\right] V_S(\rho,q) = -\frac{4\pi e^2}{\varepsilon_w}\delta(\rho) \tag{3}$$

where $q = k - k'$. The latter is the linear equation with the variable coefficient in the middle term depending on the plane coordinate ρ. A characteristic spatial change of the quantity $\partial n^{ind} / \partial \mu_0$ is significantly in the order of the CQWR radius R. As in Ref.[9] with the Q2D EG case, if replace the quantity $\partial n^{ind} / \partial \mu_0$ in equation (3) by the latter's average value such as

$$\frac{\partial \bar{n}^{ind}}{\partial \mu_0} = \frac{1}{S}\int_0^R \frac{\partial n^{ind}}{\partial \mu_0}\rho d\rho d\varphi = \frac{1}{S}\frac{\partial \bar{n}_L}{\partial \mu_0}, \tag{4}$$

we can obtain linear equation

$$\left(\frac{\partial^2}{\partial\rho^2} + \frac{1}{\rho}\frac{\partial}{\partial\rho}\right)V_S - \tilde{q}^2 V_S = -\frac{4\pi e^2}{\varepsilon_w}\delta(\rho) \tag{5}$$

with the independent of the variable ρ coefficient $\tilde{q}^2 = q^2 + q_S^2$, $\bar{n}_L$ is the mean linear charge density, $S = \pi R^2$ is the CQWR cross section area. Here in equation (5)

$$q_S = \sqrt{\frac{4\pi e^2}{\varepsilon_w}\frac{\partial \bar{n}_L}{\partial \mu_0}} \tag{6}$$

is the averaged TFA screening parameter.

As we can see from the Eq.(5) a characteristic spatial distance along the z axis in CQWR for that a solution of the equation changes appreciably, is order of the $\tilde{q}^{-1}$. If the condition $\tilde{q}R << 1$ holds then, the sought potential $V_S(\rho, q)$ from the Eq.(3) coincides with the solution of Eq.(5) and, hence, we must focus on the solution of the last equation. The most general solutions of the Eq.(5) both with the axial and reflection symmetry at the origin are [11, 12]

$$\begin{cases} V_S = \int_0^\infty \cos qz\left[AI_0(\tilde{q}\rho) + BK_0(\tilde{q}\rho)\right]dq, & \rho \le R \\ V_S = \int_0^\infty \cos qz\left[C K_0(q\rho) + D I_0(q\rho)\right]dq, & \rho > R \end{cases}, \tag{7}$$

where I(x) and K(x) are modified Bessel functions. The coefficients A(q), B(q), C(q) and D(q) have been determined by applying the standard Maxwell boundary conditions at the CQWR surface and requiring the limiting asymptotic expressions of the electrostatic potential $V_S(\rho, z)$ both at the origin and infinity. As a result, pair of the linear equations

$$\begin{cases} AI_0(\tilde{q}R) + \frac{2e}{\pi\varepsilon_w}K_0(\tilde{q}R) = CK_0(qR), \\ A\frac{\tilde{q}}{q\varepsilon_b}\varepsilon_w I_1(\tilde{q}R) - \frac{2e}{\pi q\varepsilon_b}\tilde{q}K_1(\tilde{q}R) = -C K_1(qR), \end{cases} \tag{8}$$

is derived, where the expressions D(q)=0, B(q)= - $2e^2/\pi\varepsilon_w$ are taken into account [12].

After combining solutions of Exps. (8) with the equation (7) for the sought 3D interaction screened potential of the semiconductor CQWR embedded in the dielectric barrier media obtain

$$\begin{cases} V_S = -\frac{2e^2}{\pi\varepsilon_w}\int_0^\infty dq\cos qz\left[K_0(\tilde{q}\rho) + I_0(\tilde{q}\rho)\frac{\left[\varepsilon_r\tilde{q}K_0(qR)K_1(\tilde{q}R) - qK_0(\tilde{q}R)K_1(qR)\right]}{\varepsilon_r\tilde{q}K_0(qR)I_1(\tilde{q}R) + qK_1(qR)I_0(\tilde{q}R)}\right], \\ V_S = -\frac{2e^2}{\pi\varepsilon_b}\int_0^\infty dq\cos qz\, K_0(q\rho)\frac{\tilde{q}K_0(\tilde{q}R)I_1(\tilde{q}R) + \tilde{q}K_1(\tilde{q}R)I_0(\tilde{q}R)}{\varepsilon_r\tilde{q}K_0(qR)I_1(\tilde{q}R) + qK_1(qR)I_0(\tilde{q}R)}. \end{cases} \tag{9}$$

where $\varepsilon_r = \varepsilon_w/\varepsilon_b$. As follows, the screened potential in the QWR region consists of the homogeneous and inhomogeneous polarization connected parts such as: $V_S = V_S^{hom} + V_S^{inh}$. In the case of an infinite wire the potential inside the wire takes the fully screened 3D Debye-Hückel potential form as $R \to \infty$

$$V_S(\rho, z) = -\frac{e^2}{\varepsilon_w}\frac{e^{-q_S\sqrt{\rho^2+z^2}}}{\sqrt{\rho^2+z^2}}. \tag{10}$$

For the case of infinitesimal wire or filament ($R \to 0$) the screened potential outside the QWR takes the unscreened Coulomb potential form characterized by barrier dielectric constant

$$V(\rho, z) = -\frac{e^2}{\varepsilon_b}\frac{1}{\sqrt{\rho^2 + z^2}}. \tag{11}$$

Note, that the equation (9) easily reduces to the appropriate unscreened CQWR 3D interaction potential forms (see Ref.[16]) if neglect the TFA screening parameter q_S in the effective Q1D wave vector expression $\tilde{q} = \sqrt{q^2 + q_S^2}$. In turn, the screened potential would be reduced as well to the metallic CQWR case expressions (Refs.[11, 12]) by substitution $\varepsilon_w = \varepsilon_b = 1$.

3. Dielectric confinement affected Q1D CQWR screening potential and effective screening radius

Now we will obtain CQWR Q1D interaction screened potential analytic expressions taking into account the DC effect. In order to get a general insight into the "Coulomb interaction engineering" possibilities of the DC effect a strong contrast between the dielectric constants ε_w and ε_b is accepted such as: $\varepsilon_r = \varepsilon_w / \varepsilon_b >> 1$.

In the previous section the condition $\tilde{q}R << 1$ has been imposed meaning that the typical spatial change of the screened potential energy's Fourier-component is large enough than the CQWR radius. The latter will be realized when the following two conditions are simultaneously fulfilled

$$\begin{cases} qR << 1, \\ q_S R << 1. \end{cases} \tag{12}$$

With these conditions the screened potential takes the Q1D form such as

$$V_S^{Q1D}(z) = -\frac{e^2}{\varepsilon_w}\frac{e^{-q_S|z|}}{|z|} - \frac{2e^2}{\pi\varepsilon_w}\int_0^\infty \cos qz \frac{\varepsilon_r \ln(qR)^{-1}}{\varepsilon_r(\tilde{q}R)^2 \ln(qR)^{-1} + 1} dq. \tag{13}$$

If we neglect the DC effect, i.e. the condition $\varepsilon_r \sim 1$ would be imposed in equation (13), then 1D screened potential long wave limit result has to be restored (Exps.4.3-4.4 in Ref. [14]) in QWR.

As for the unscreened case in QWR [16, 18], for the asymptotic distances along the wire axis $|z| >> R$ there exists in equation (13) two distinct effective long wave vector ranges appropriate to the 1D moderate small and very small wave vectors. The latters are: $\varepsilon_r (qR)^2 \ln(qR)^{-1} \geq 1$ and $\varepsilon_r (qR)^2 \ln(qR)^{-1} << 1$, respectively, for that the Q1D screened potential takes the following asymptotic analytic forms.

3.1. A

For the intermediate distances $|z| >> R$ corresponding to the criteria

$$\varepsilon_r \left(\frac{R}{z}\right)^2 \ln\frac{|z|}{R} \geq 1, \tag{14}$$

after identical transformations connected with the first wave vector inequality, the screened interaction potential takes the form

$$V_{Si}^{Q1D}(z) = -\frac{e^2}{\varepsilon_w}\left[\frac{e^{-q_S|z|}}{|z|} - \frac{2}{\pi R^2}\int_0^\infty \frac{\cos qz dq}{q^2 + \tilde{q}_S^2}\right]. \tag{15}$$

The latter, in turn, gives the result as

$$V_{Si}^{Q1D}(z) = -\frac{e^2}{\varepsilon_w R}\left[\frac{R}{|z|}e^{-q_S z} + \frac{1}{\tilde{q}_S R}e^{-\tilde{q}_S z}\right], \tag{16}$$

where

$$\tilde{q}_S = \frac{1}{R}\sqrt{\frac{2}{\varepsilon_r \ln \varepsilon_r} + q_S^2} \ . \tag{17}$$

From the standpoint of both long wave ($q << R^{-1}$) and DC effect ($\varepsilon_r = \varepsilon_w / \varepsilon_b >> 1$) conditions, the main contribution in equation (16) connected with the inhomogeneous polarization related second term.

Afterwards, the screened interaction potential will take the final form

$$V_{Si}^{Q1D}(z) = -\frac{e^2}{\varepsilon_w R}\frac{1}{\tilde{q}_S R} e^{-\tilde{q}_S z} \ . \tag{18}$$

As can be seen from equation (18) the DC affected screening potential deviates from the 3D and pure 1D screened potential forms. It takes the Q1D form and, as in the Q2D case [9], strongly enhances with the reducing of the QWR confinement length. The similar result is established in Ref.[14] as well, where the DC effect is out of the discussion.

Note, that the Q1D screening effect emerges for the moderately large $|z| >> R$ intercharge distances only, which is appropriate to the moderately thin quantum wires. The screened potential $V_{Si}^{Q1D}(z)$ strongly depends from the DC affected effective screening parameter $\tilde{q}_S$ and, in turn, is enhanced due to the small parameter $\tilde{q}_S R << 1$. At the same time, $V_{Si}^{Q1D}(z)$ is exponentially weak enough outside of the spatial region with the linear size equal to $z_S = \tilde{q}_S^{-1}$, which is believed to be the effective Q1D screening distance in CQWR. Thereby takes place a recovering of the effective Q1D screening length, like a DC affected QW case [9, 25]. Besides that, the Q1D screening length z_S strongly depends on the dielectric constants ratio ε_r. As follows, accounting of the DC effect results that the screening effect becomes effective in this case.

Let now study the TFA screening averaged parameter after equation (6).

Within the framework of the TFA method it is believed that only one size-quantized energy subband is filled (the size quantum limit (SQL)) and that the spatial variation of the electrostatic potential is small over distances comparable to the mean electron de Broglie wavelength λ_{dB}. So in SQL the QWR is believed thin enough (R < λ_{dB}) that the long wave condition qR< 1 after equation (12) holds. In turn, the mean linear charge density $\bar{n}_L$ after equation (6) is determined by 1D EG statistics at 0K as [24, 14]

$$n_L = \frac{4}{\pi \hbar}\sqrt{\frac{m_e}{2} E_F} \ , \tag{19}$$

where $E_F = \hbar^2 k_F^2 / 2m_e$ is the Fermi energy of the 1D EG and the Fermi wave number is determined as $k_F = \pi n_L / 2$, m_e is the electron effective mass. If combining the Exps. (6) and (19) we get

$$q_S = \frac{4}{\pi R}\left[\frac{1}{a_0 n_L}\right]^{1/2} . \tag{20}$$

and

$$z_S = R\left[\frac{2}{\varepsilon_r \ln \varepsilon_r} + \frac{16}{\pi^2 a_0 n_L}\right]^{-1/2} \tag{21}$$

The second inequality in (12) in accordance with equation (20) results that $a_0 n_L >> 1$, i.e. the number of 1D electrons on the Bohr radius length should be large.

3.2. B

For the very large intercharge distances $|z| >> R$ corresponding to criteria

$$\varepsilon_r (\frac{R}{z})^2 \ln \frac{|z|}{R} << 1 \quad (22)$$

and with the small enough related values of q_S such as $\varepsilon_r (q_S R)^2 \ln(qR)^{-1} << 1$, the screened potential with the second wave vector inequality goes to the form

$$V_{Sii}^{Q1D}(z) = -\frac{e^2}{\varepsilon_b \sqrt{z^2 + R^2}} . \quad (23)$$

Equation (23) is appropriate to the extremely thin wires and illustrates the following features.

At first, the latter reduces to the 1D Coulomb screened potential when DC effect is absent (see equation (4.5) of Ref.[14]). In turn, $V_{Sii}^{Q1D}(z)$ is characterized, as in the unscreened case [16], by Q1D Coulomb law with the surrounding barrier environment small value dielectric constant. Thus, in the case of thin enough quantum wires, the Q1D screening effect is wholly suppressed regardless of the accounting the DC effect. That is fairly natural, since in the Q1D (1D) systems for the large enough intercharge distances z, the predominant part of the electric field lines extend outside the free carrier confinement region. As a result, the free carriers cannot effectively screen the Coulomb field between any two charges inside the thin enough quantum wire.

4. Conclusion

In summary, the asymptotic relatively simple expressions of the screened Coulomb interaction potential and effective screening length of the semiconductor CQWR embedded in the dielectric barrier media is derived. It is shown, that for the moderately thin DC influenced semiconductor CQWR the screened Q1D interaction potential to be

$$V_{Si}^{Q1D}(z) = -\frac{e^2}{\varepsilon_w R} \frac{1}{\tilde{q}_S R} e^{-\tilde{q}_S z} . \quad (24)$$

The latter becomes strongly enhanced when reducing both wire radius and the Q1D effective screening parameter $\tilde{q}_S$. Expression of the DC affected effective Q1D screening length

$$z_S = \frac{R}{\left[\frac{2}{\varepsilon_r \ln \varepsilon_r} + \frac{16}{\pi a_0 n_L} \right]^{1/2}} \quad (25)$$

is established in the TFA approximation, which strongly depends on the dielectric constants ratio ε_r. It is received, that, like a DC affected QW case, in DC influenced QWR system takes place a recovering of the effective Q1D screening length. It has been revealed as well that the screening effect does wholly diminish in thin enough quantum wire, in accordance with Q1D Coulomb law (23). The latter is characterized by barrier environment small dielectric constant as in the unscreened case.

Acknowledgment

The authors are thankful to Professor E.M. Kazaryan for the helpful discussions.

References

[1] Timp G (ed) 1999 *Nanotechnology* (New York: Springer-Verlag)
[2] Wang X L and Voliotis V 2006 *J. Appl. Phys.* **99** 121301
[3] Koh W-k, Bartnik A C, Wise F W and Murray C B 2010 *J. Amer. Chem. Soc.* **132** 3909
[4] Derycke V, Martel R, Appenzellet J and Avouris P 2001 *Nano Lett.* **1** 453
[5] Li Y, Qian F, Xiang J and Lieber C M 2006 *Materials Today* **9** 18

[6] Pinto F 2013 *J. Appl. Phys.* 2013 **114** 024310
[7] Gubarev S I, Kukushkin I V, Tovstonog S V, Akimov M Yu, Smet J, von Klitzing K and Wegschelder W 2000 *JETP Lett.* **72** 324
[8] Ashcroft N V and Mermin N D 1976 *Solid State Physics* (Philadelpia: Saunders College)
[9] Rytova N S 1967 *Sov. Phys. Doklady* **10** 754; 1967 *Proc. MSU*, Physics, Astronomy **3** 30
[10] Stern F 1967 *Phys. Rev. Lett.* **18** 546
[11] Kuper C G 1966 *Phys. Rev.* **150** 189
[12] Davis D 1973 *Phys.Rev.* B **7** 129
[13] Lee J and Spector H N 1985 *J. Appl. Phys.* **57** 366
[14] Reyes J A and del Castillo-Mussot M. *Phys. Rev.* B **57** 9869
[15] Wang Y, Miao W-D and.Zhai L-X 2014 *Phys. Lett.* A **378** 442
[16] Babichenko V S, Keldysh LV and Silin A P 1980 *Sov.Phys.Solid State* **22** 1238; Keldysh LV 1997 *Phys. St. Sol.* (a) **164** 3
[17] Aharonyan K H and Kazaryan E M 1982 *Sov. Phys. Semicond.* 16 122; 1983 *Sov. Phys. Semicond.* **17** 1140
[18] Shik A 1993 *J. Appl. Phys.* **74** 2951
[19] Wendler L and Grigoryan V G 1994 *Phys. Rev.* B **49** 14531
[20] Aharonyan K H, Erknapetyan H L and Tilley D R 1988 *Phys. St. Sol.* (b) **150** 133
[21] Aharonyan K H and Khachatryan A Zh 2004 *Proc. of SEUA Annual conf., Yerevan* **1** 46
[22] Konar A, Fang T and Jena D 2011 *Phys.Rev.* B **84** 085422
[23] Bartnik A C, Efros Al. L, Koh W-K, Murray C B and Wise F W 2010 *Phys. Rev.* B **82** 195313
[24] Celina E, Magana F, Valladares A A 1977 *Amer. J. Phys.* **45** 960
[25] Aharonyan K H and Kazaryan E M 2012 *Physica E.* **44** 1924

Design and development of a LabVIEW-based LED-induced fluorescence spectroscopy system with applications in non-destructive quality assessment of agricultural products

Hamed Abbasi[1], Majid Nazeri[1] and Seyed Ahmad Mireei[2]

[1] Department of Photonics, Faculty of Physics, University of Kashan, Kashan 87317-51167, Iran

[2] Department of Biosystems Engineering, College of Agriculture, Isfahan University of Technology, Isfahan 84156-83111, Iran

Email: hamed.abbasi@grad.kashanu.ac.ir

Abstract. Over the past several years, the demand for high quality agricultural products has been remarkably increased. Thus, it is important to use non-destructive methods for product quality monitoring. LED-induced fluorescence spectroscopy has proved its potential for non-destructive detection of some defects in agricultural products, such as tissue browning and bruising. Due to such defects, changes in the polyphenol and chlorophyll contents occur which can be considered as the visible marks of decreasing fruit quality. In the present work, a fluorescence spectrometer (spectrofluorometer) controlled by LabVIEW software was designed and developed. In this spectrometer, a consumer-grade webcam was used as an imaging sensor. The spectrometer was able to measure the fluorescence spectra directly from the fruit and vegetable surface in the desired regions. To do so, the spectrometer was equipped with a suitable fiber-optic probe. The hardware solution was based on data acquisition working on the USB platform and controlled by the application running on the PC. In this system, light emitting diodes with different wavelengths were used as the excitation sources for inducing fluorescence spectra of some famous fruits and vegetables.

1. Introduction

Nowadays, spectral measuring methods such as fluorescence, transmission, remission and diffuse reflection have been investigated for fruit quality control applications [1]. Fluorescence spectroscopy has proved its potential for various applications in agriculture, such as detection of tissue browning and bruising [2-4] and freshness control [5]. Also, it can be used to monitor water stress in plants [6]. Due to such defects, changes in molecular structure of the chlorophyll of products occur which can be considered as visible marks of decreasing fruit quality. Plants show a fluorescence emission in the visible spectral range that is excited by UV radiation and partially also by visible light [1]. The chlorophyll fluorescence of agricultural products mainly consists of two maxima in the red and far-red region. The intensity and shape of the chlorophyll fluorescence emission spectrum at room temperature are primarily dependent on the concentration of the fluorophore chlorophyll-α, and to a lower degree also on the structure, the photosynthetic activity, and the it's characteristics and arrangements of cells in the tissue. [7]. A Laser-induced fluorescence (LIF) spectroscopy system consists of two main parts: a sensitive spectrometer and a source for excitation. It is possible to use different sources to induce the fluorescence in fruits and vegetables [8-10]. In order to select an appropriate excitation source, it is very important to know the relation between the excitation source wavelength and the emission wavelength [11]. In this study, the power light emitting diodes (LEDs) rather than laser have been used as the excitation source, due to their lower price. There some attempts for developing cost-effective LIF spectroscopy system with application in agriculture sciences. Paul K Buah-Bassuah et al. developed a portable fibre-probe ultraviolet light emitting diode (LED)-induced fluorescence detection system by combining an Ocean Optics spectrometer and a 365 nm UV LED [12]. In this paper, first, a home-made spectrometer using a consumer-grade webcam was designed and developed. In the second step, the fluorescence spectra of the fruits and vegetables were collected with the five different excitation sources, including power UV and blue light emitting diodes (LEDs) with different wavelengths.

2. Materials and methods

In this study a prism-based spectrometer (which has the higher sensitivity compared to grating-based spectrometers) controlled by LabVIEW was developed in order to acquire the spectra in the interval of 571 to 1149 nm. Figure 1 shows the schematic diagram of the spectrometer.

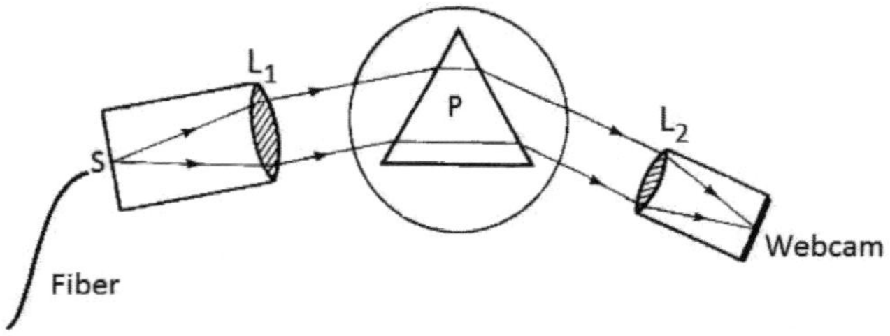

Figure 1. The schematic diagram of the spectrometer. “S”, “P”, “L1” and “L2” represent slit, prism, collimator lens and the imaging lens (camera lens), respectively.

The entrance light illuminates the slit, which was placed in the focal plane of the collimator lens. Behind the collimator lens, the parallel light beam passes through the prism, where it is diffracted depending on the wavelength λ. The imaging lens forms an image of the entrance slit on the webcam sensor. A consumer-grade webcam (Dell, 640 X 480 pixels) was used in this study. The position of the image of slit in the focal plane of imaging lens was a function of the wavelength λ. Figures 2 and 3 show the internal and external schematic of the developed spectrometer, respectively.

Figure 2. Internal schematic of the developed spectrometer.

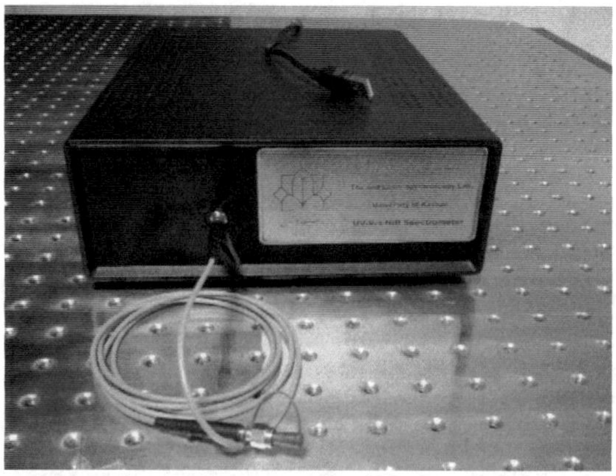

Figure 3. External schematic of the developed spectrometer.

The first wavelength in the free spectral range (FSR) of the spectrometer was 571 nm, and the last one was 1149 nm. The triangle flint prism was used in this spectrometer. The prism was used at the minimum angle of deviation. In this angle, the wavelength of interest traveled parallel to the base of the prism, and the angle of incidence was equal to the angle of refraction. The refraction index of the used prism was 1.6216, 1.6207, 1.6005 and 1.6003 in the wavelengths of 571, 581, 1139 and 1149 nm, respectively. Therefore, for the first and last 10 nm of FSR, the spectral dispersion, dn/dλ, equals:

$$\frac{dn}{d\lambda} = \frac{1.6207 - 1.6216}{10\,\text{nm}} = -0.9 \times 10^{-4}\,\text{nm}^{-1} \quad (1)$$

$$\frac{dn}{d\lambda} = \frac{1.6003 - 1.6005}{10\,\text{nm}} = -0.2 \times 10^{-4}\,\text{nm}^{-1} \quad (2)$$

The spectral resolving power, R, of any triangle prism is defined as follows [13]:

$$R = \frac{\lambda}{\Delta\lambda} \leq \frac{1}{3} g \left(\frac{dn}{d\lambda}\right) \quad (3)$$

In this formula "g" is base length of the prism. In this study, the base length of the prism was 25 mm, therefore for the first 10 nm of the spectrometer, the maximum value of R can be calculated as 750. Also, the smallest resolvable wavelength interval, Δλ, in 571 nm was calculated as follows:

$$\Delta\lambda_{\min} = \frac{571\,\text{nm}}{750} = 0.76\,\text{nm} \quad (4)$$

With similar method, the maximum value of R for the last 10 nm of the spectrometer and the smallest resolvable wavelength interval, Δλ, in 1149 nm were calculated as 166.6 and 6.8 nm, respectively.

Obviously, the operation of the spectrometer in the visible region was better than the near infrared region. The spectrometer was able to collect the fluorescence spectra directly from the fruit and vegetable surface in the desired regions. To do so, the spectrometer was equipped with a suitable fiber-optic probe. The hardware solution was based on the data acquisition working on the USB platform and controlled by the application running on the PC.

Generally, in prism spectrometers, the resolution varied at the different wavelengths. Therefore, a number of reference lines are required for calibration. In this study, lines of Mercury (Hg), Sodium (Na) and Cadmium (Cd) lamps were used as well as the first and second harmonic of Nd-YAG laser for calibration of the spectrometer. Figure 4 shows the lines of the Mercury lamps on the webcam sensor, in case of adjusting the spectrometer from 360 to 1100 nm.

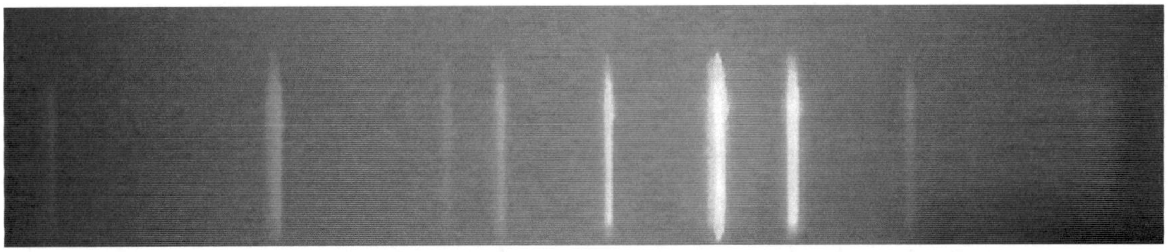

Figure 4. Lines of the Mercury lamps on the webcam sensor (360-1100 nm).

We improved the dynamic range of our system by means of using a large number of rows of webcam sensor (instead of using a single row of that) without going to saturation (by trying to have a homogenous vertical distribution of spectral lines). The webcam had a 640 pixel in length. Usually, the prism spectrometers are calibrated either by using the famous formula such as Sellmeier or Cauchy's equation [14] or applying a suitable formula that are able to fit in desired region. Figure 5 shows the fitted formula for calibration of the spectrometer (360-1100 nm) with R-squared (0 to 1) value very close to 1. R-squared is a statistical parameter of how close the data are to the fitted regression line which normally is defined between 0 to 1 or 0 to 100.

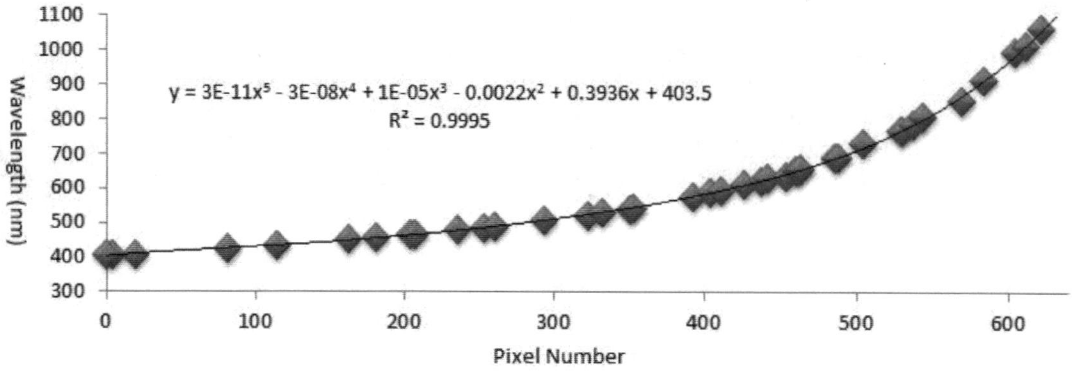

Figure 5. The trendline for calibration of the spectrometer with high R-squared value. The horizontal axis represents the pixel number and the vertical axis represents wavelength in nm.

LabVIEW software was employed to control the system. Figure 6 shows a print screen of the user interface of the LabVIEW program.

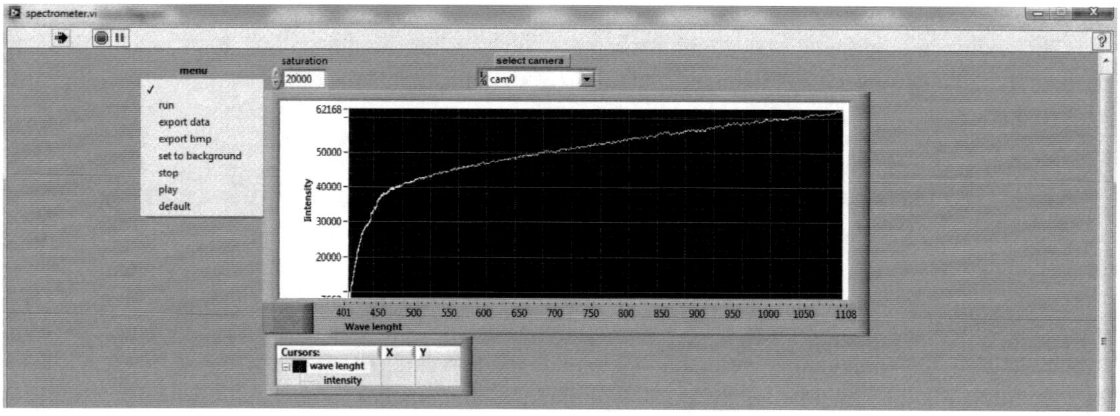

Figure 6. Print screen of the user interface of the LabVIEW program.

In LIF Spectroscopy, lasers are mainly used as the excitation source. In recent years, by developing the technology of producing LEDs, they have gradually taken the place of lasers in LIF spectroscopy. Due to the Kasha's rule, the fruits and vegetables have the excitation-wavelength-dependent fluorescence behaviour, and consequently, all of the used LEDs behave in the similar way [11]. Figure 7 shows the spectrum of the used LEDs.

Journal of Physics: Conference Series **672** (2016) 012010 doi:10.1088/1742-6596/672/1/012010

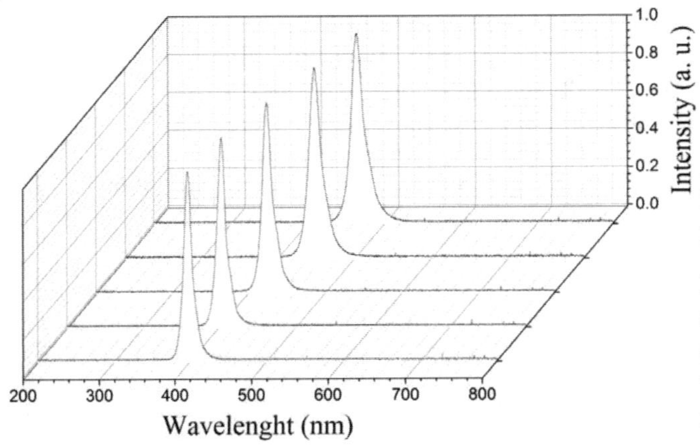

Figure 7. The spectrum of the used LEDs with the similar power values of 1 Watt and the central wavelengths at 397.5 nm, 403.5 nm, 424 nm, 448 nm, and 465.5 nm. The horizontal axis represents the wavelength in nm and the vertical axis represents the normalized intensity in arbitrary unit.

3. Results

One of the most important factors in designing and developing a laser or LED-induced fluorescence spectroscopy system for quality assessment of agricultural products is the capability of the instrument to induce and collect the fluorescence spectra of various types of fruits and vegetables. Therefore, it was necessary to test the system for various types of fruits and vegetables. Since the fruits and vegetables have the excitation-wavelength-dependent fluorescence behavior, it is possible to use the mentioned sources as the excitation source in LED-induced fluorescence spectroscopy. Figure 8 shows the fluorescence spectra of some famous fruits and vegetables including pear, peach, cherry, orange, red apple, yellow apple, water melon, carrot, potato, and basil.

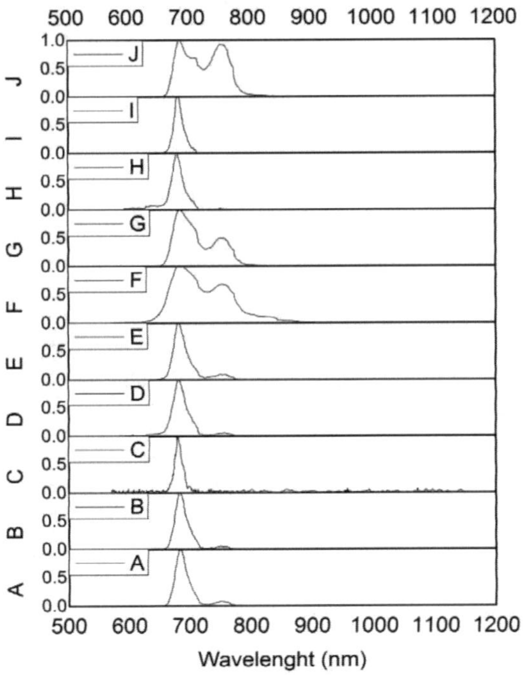

Figure 8. The collected fluorescence spectra of some famous fruits and vegetables. The first row, A, the second row, B, …, and the last row, J, represent the fluorescence spectrum of pear, peach, cherry, orange, red apple, yellow apple, water melon, carrot, potato, and basil respectively.

For the basil, water melon and Golden Delicious apple, three peaks were observed at 690 nm, 735 nm and 792 nm. The peaks at 690 nm (red fluorescence) and 735 nm (far red fluorescence) are related to the chlorophyll-α fluorescence and are the good indices to assess the chlorophyll-α concentration. The variation in the peak of 740 nm can be due to the inactive photosynthesis process [12, 15]. It should be also be noted that the present of fluorophore in the basil leaf and other fruits which are found by the 690 nm and 740 nm fluorescence peaks, can be used to explain the chlorophyll degradation during the photosynthesis. The peak at 792 nm may be due to the unwanted wavelength of laser-light other wavelength than expected. For the other fruits and vegetables, including the Red Delicious apple, orange, peach, pear, potato, carrot and cherry, the peaks at 735 nm and 792 nm diminished drastically and only a strong peak at 690 nm can be observed. This can be due to the fact that the samples of this group had the lower active photosynthesis process and their chlorophyll status were significantly less than the first group.

4. Conclusion

In this paper, a spectrometer was designed and developed using commercial components which was controlled by the LabVIEW software and equipped with an appropriate fiber-optic probe. The 200 micrometer fiber optic was used as probe; the triangle prism was used as a diffraction element; and the consumer-grade webcam was applied as the imaging sensor. The lines of Mercury (Hg), Sodium (Na) and Cadmium (Cd) lamps as well as the first and second harmonic of Nd-YAG laser were used for calibration of the spectrometer. The light emitting diodes (LEDs) with different wavelengths in UV and blue regions were selected as the excitation source We successfully tested our system for different types of fruits and vegetables to make sure that it was able to work correctly, because different types of agricultural products have different amount of fluorescence emission and one system may be able to collect the fluorescence of one type of vegetable/fruit but not capable of collect the fluorescence spectra of another vegetable/fruit (because it may need another excitation source or more sensitive spectrometer).

Acknowledgments

We thank Mr. Mohammad Reza Nazeri for his valuable helps and comments in the LabVIEW programing.

References

[1] Zude M 2008 *Optical monitoring of fresh and processed agricultural crops* (New York: CRC Press)
[2] Wulf J S, Geyer M, Nicolai B and Zude M 2004 *Proc. 5th Int. Postharvest Symp.* (Verona: Italy) pp 1387-94
[3] Zude M 2002 *Proc. XXVI Int. Horticultural Cong.: Issues and Advances in Postharvest Horticulture* (Toronto: Canada) pp 85-91
[4] Wulf J S, Werne B H, Geyer M and Zude M 2003 *Proc. Int. Conf. on Quality in Chains. An Integrated View on Fruit and Vegetable Quality* (Wageningen: Netherlands) pp 653-658
[5] Abbasi A, Bani M A, Nazeri M and Mireei S A 2014 Application of Fluorescence Spectroscopy in the Freshness Control of Lettuce Leaf after Irradiation of Cold Atmospheric Pressure Plasma *Proc. 2nd Conf. on Plasma Engineering and Physics* (Babolsar: Iran)
[6] Lins E C, Nunes F D, Gasparoto M C G, Junior J B, Bagnato V S and Marcassa L G 2005 *Proc. IEEE Int. Conf. on Microwave and Optoelectronics, SBMO/IEEE MTT-S* (Brasilia: Brazil) pp 534-537
[7] Buschmann C, 2007 *Photosynth Res* **92** 261
[8] Moya I, Camenen L, Evain S, Goulas Y, Cerovic Z G, Latouche G, Flexas J and Ounis A 2004 *Remote Sens. of Environ.* **91** 186
[9] Da Silva A J, do Nascimento C W A, da Silva Gouveia-Neto A and da Silva-Jr E A 2012 *Water Air Soil Pollut.* **223** 3527
[10] De Wilton A C and Koningstein J A 1985 *Chem. Phys. Lett.* **114** 161
[11] Abbasi H, Nazeri M, Mireei S A and Balooch A 2014 LIF Spectroscopy of Fruits: Study of Excitation Wavelength Independence *Proc. 3rd Iranian Conf. on Engineering Electromagnetic* (Tehran: Iran)

[12] Buah-Bassuah P K, von Bergmann H M, Tatchie E T and Steenkamp C M 2008 *Meas. Sci. Technol.* **19** 025601
[13] Demtröder W 2014 *Laser Spectroscopy 1: Basic Principles* (Berlin Heidelberg: Springer-Verlag)
[14] Pedrotti F L and Pedrotti L S 1993 *Introduction to Optics* (New Jersey: Prentice Hall)
[15] Liu C, Song Y, Zhang D, Zhang, H 2006 *Proc. IEEE Int. Symp. on Biophotonics, Nanophotonics and Metamaterials* (Hangzhou: China)

Journal of Physics: Conference Series **672** (2016) 012011 doi:10.1088/1742-6596/672/1/012011

Coherence modulation at the photon-counting level: A new scheme for secure communication

William T Rhodes[1,3], Abdellatif Boughanmi[1] and Yezid Torres Moreno[2]

[1] Florida Atlantic University, Department of Computer & Electrical Engineering and Computer Science, Boca Raton, Florida 33431, USA
[2] Universidad Industrial de Santander, GOTS, Grupo de Óptica y Tratamiento de Señales, Escuela de Fisica, Facultad de Ciencias, Bucaramanga, Colombia

E-mail: wrhodes@fau.edu, latif2178@yahoo.com, ytorres@uis.edu.co

Abstract. When operated at the photon-counting level, coherence modulation can provide quantifiably secure binary signal transmission between two entities, security being based on the nonclonability of photons.

1. Introduction

The secure communication paradigm is illustrated in Figure 1. Alice and Bob want to communicate in complete secrecy; Eve wants to "listen in" on their communication. Three standard methods are identified for secure communication: (a) steganography, where the message is disguised as something else; (b) system hardening, where the communication link is made physically inaccessible; and (c) cryptography, where the message is encrypted in such a way that it is meaningless without the key. In this paper we consider a forth method: transmit the encrypted message in such a way that Bob can receive it correctly, but if Eve attempts to intercept it, it takes the form of random noise.

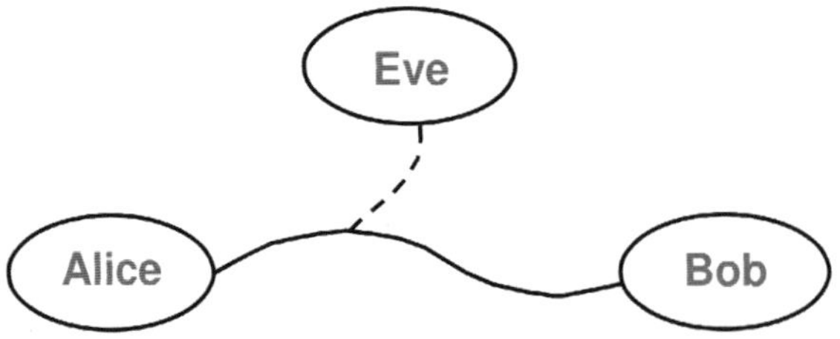

Figure 1. The secure communication paradigm.

The case of conventional cryptography is illustrated in Figure 2. In order to "break" the code, Eve must have access to the ciphertext. In our scheme, we prevent Eve from accessing the ciphertext by physical processes that—for her but not for Bob—convert it to a string of random 1's and 0's. The underlying method is referred to as key-specified coherence modulation, operating in our case at very low, photon-counting, light levels. As we will show, Bob, knowing the transmission key, can receive the message virtually error-free. Eve, not knowing the transmission key, detects a succession of

[3] Corresponding author.

random 1's and 0's. Furthermore, in tapping into the low-intensity light transmitted by Alice to Bob, Eve betrays her presence. In contrast to the case of conventional cryptography, the security of the transmission is guaranteed by laws of physics—by the non-clonability of individual photons—rather than by wished-for unbreakability of a particular encryption scheme. Clearly the scheme has much in common with quantum key distribution, a secure communication scheme that has received considerable attention during the past several decades [1].

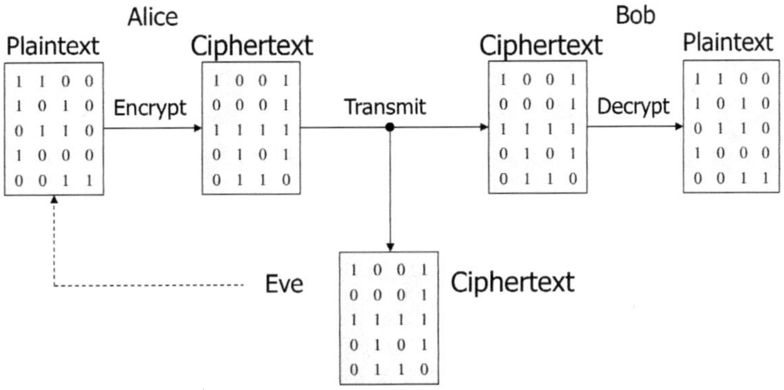

Figure 2. Conventional cryptography. To break the code Eve must have access to the ciphertext. Our scheme prevents Eve's this access; instead, she sees only random 1's and 0's.

Before continuing, we recall two basic principles of cryptography. First, it is assumed that Eve knows the encryption algorithm. Second, it is also assumed that Eve is extraordinarily capable: she has the best code breakers, the best possible science, absolutely outstanding technology, etc. A corollary is that security lies in having a strong key and in keeping that key secret (in combination with a good algorithm, of course).

The remainder of the paper is organized as follows: Section 2 describes how 1's and 0's can be communicated by means of zero-pi phase shifting in an interferometer. In Section 3 coherence modulation for "secure" communication is introduced. Section 4 describes a possible high-light-level attack by Eve that gives her access to the ciphertext. Section 5 shows how operation at the photon-counting level precludes a successful attack by Eve. A specific example of a statistical analysis is presented in Section 6, and concluding remarks are made in Section 7.

2. Communicating 1's and 0's by a phase-shifting interferometer

Figure 3 shows a Mach-Zehnder interferometer with a phase shifter in one arm that allows the phase ϕ of the light in that arm to be switched (achromatically, i.e., for all wavelengths) by zero or pi radians. Switching ϕ conveys the value of the bit to be communicated: if $\phi = 0$, signifying a 1-bit, all the light exits the interferometer to the right; if $\phi = \pi$, signifying a 0-bit, all the light exits to the top. Although the figure shows laser light input, the device works with white light too, so long as the phase shift ϕ is independent of wavelength and so long as the path length difference for the two arms of the interferometer is small compared to the coherence length of the light.

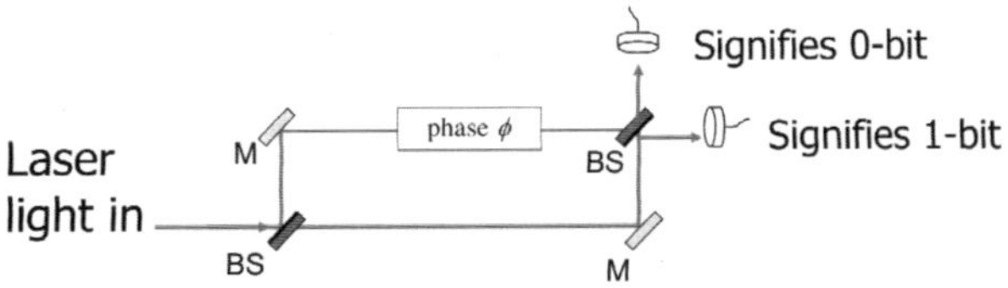

Figure 3. Phase shifting Mach-Zehnder interferometer. All the light exits to the right if $\phi = 0$, signifying a 1-bit. If $\phi = \pi$, all the light exits to the top, signifying a 0-bit.

3. Coherence modulation for "secure" communication

Figure 4 shows the phase shifting interferometer but now with the addition of two time delays, one controlled by Alice (T_A, top arm), the other controlled by Bob (T_B, bottom arm). If T_A equals T_B, phase switching still works, even for white light. However, if $|T_A-T_B| >> \tau_c$, the coherence time of the light, there is no interference; light goes to both outputs in equal amounts.

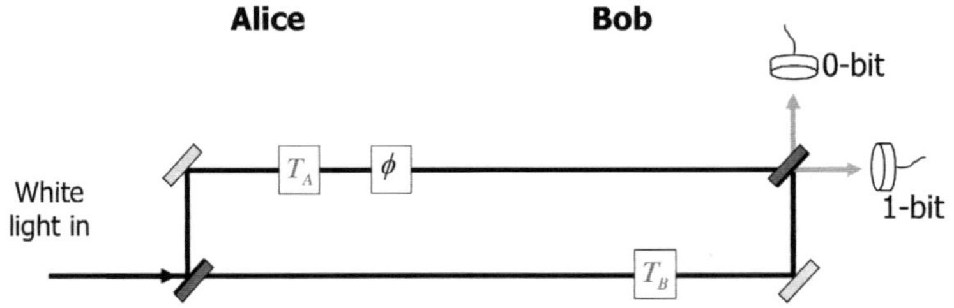

Figure 4. Phase shifting interferometer with time delays T_A and T_B. If $|T_A-T_B|$ is sufficiently large compared to the coherence length, the light exits both ports in equal amounts.

The idea underlying "secure" communication via coherence modulation (the word secure is in quotes, because, as explained below, the scheme is, in fact, quite insecure) is thus the following [2]:

- Alice and Bob share a key in the form of the delay time sequence.
- They both use the same delays in synchronism.
- The set of possible time delays is drawn from a set of M different delays.
- They keep the time delay sequence secret from Eve.

4. The problem with coherence modulation at high light levels: A possible attack by Eve

Since Eve does not know the sequence of delays, it would appear that she cannot receive the information transmitted by Alice. In fact, however, the system is not secure. Different possible attacks have been identified. In order for Eve to be successful with these attacks, it is only necessary that the light intensities at the output ports be sufficiently high, i.e., in a "classical" or non-photon-counting regime. One possible attack is illustrated in Figure 5. In this attack, Eve has M "receivers" operating in parallel, each corresponding to one of the M possible time delays used by Alice and Bob (recall, it is assumed that Eve knows the algorithm, is lacking only the key).

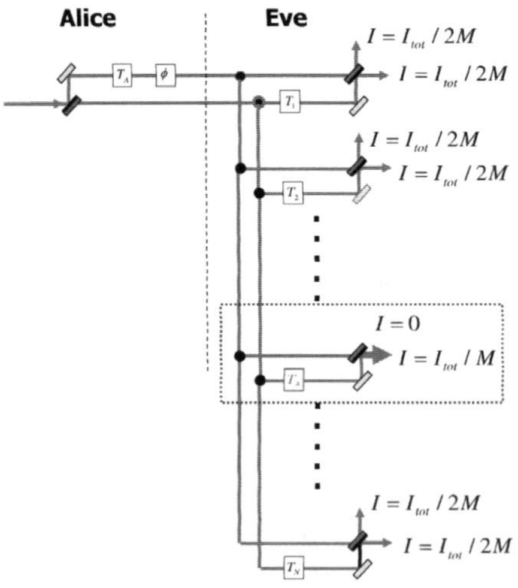

Figure 5. One possible attack by Eve. Eve splits light equally among M interferometers. The output at one port will be zero, signifying both the time delay and the bit value.

In this attack, the light is split equally between all "receivers," one for each of the M possible time delays. One of the output ports has zero output, telling Eve which time delay T_A was used by Alice. Eve thus knows both the time delay T_A and the bit, 0 or 1. Eve can send light on to Bob with the correct properties, thereby hiding her presence. As noted, this attack requires that the light in the system be at high intensity levels. As discussed in the next section, operation by Alice and Bob at the photon-counting level can prevent Eve from gaining such information.

5. How operation at the photon-counting level precludes a successful attack by Eve

We begin with a key observation regarding operation at photon-counting light levels. As suggested in Figure 6, in going from the classical light level case (top) to the photon counting case (bottom), a given light intensity is replaced with the probability of a photon count. Specifically, when the output light intensity is reduced to the level of photon counting, the *probability* of detecting a photon at a particular output port is proportional to the classical intensity of the light at that port. In the figure, ideal operation is assumed. Switching ϕ has the same result even if light is attenuated to the photon-counting level. Figure 7 illustrates the bit-transmission scheme at photo-count levels: $\phi = 0$ produces a photo-count at one output port, whereas $\phi = \pi$ produces a photo-count at the other port.

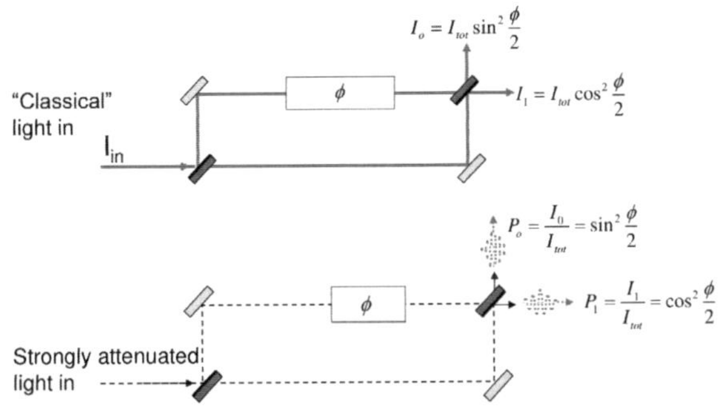

Figure 6. Reducing light to photon-counting level. Light intensities are replaced by photo-count probabilities.

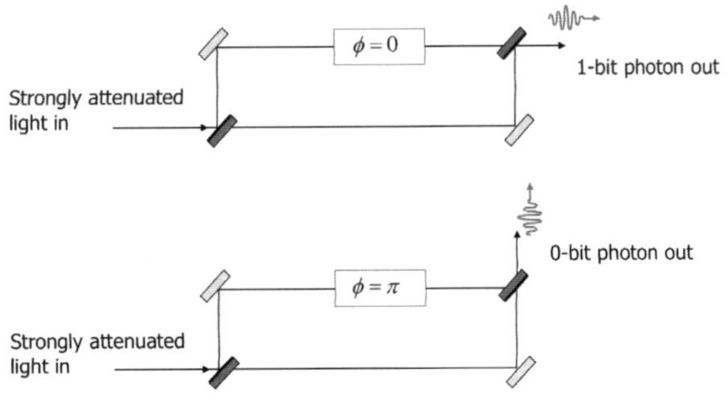

Figure 7. Interferometer operation at the single photon level.

It is assumed in figures Figure 6 and Figure 7 that the interferometer is adjusted for equal time delays in the upper and lower arms. With unequal time delays, things are different, as suggested in Figure 8. Specifically, if $|T_A - T_B| \gg \tau_c$, where τ_c is the coherence time of the light, photo-counts occur with equal probability at both output ports. No information is conveyed.

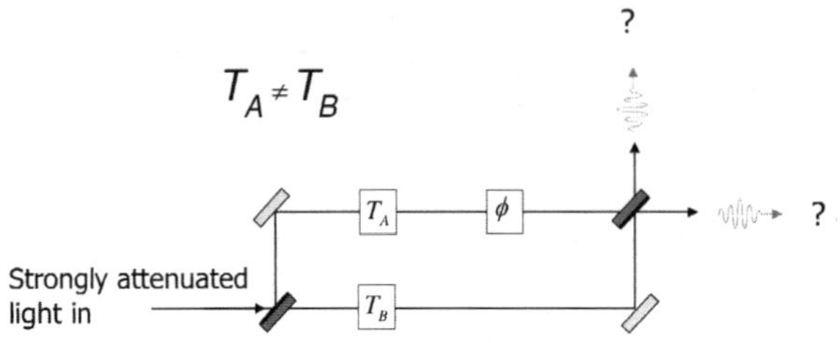

Figure 8. Single photon level with $|T_A - T_B| \gg \tau_c$. Photo-counts are recorded at both outputs with equal probability.

Is it possible for Eve to attack the system in the low-light-level regime? With high light levels (classical light) it was easy for Eve to perform many interference experiments simultaneously, in parallel, using all possible time delays, and to determine the time delay used by Alice, along with the associated bit value. Unfortunately (for Eve), the exhaustive parallel-detection attack does not work at the photon-counting level with small numbers of photons. In that case, Eve gains virtually no information. Figure 9 shows Eve's system in the photon-counting regime. As before, classical light intensities are replaced by photo-count probabilities, unit quantum efficiency being assumed. With single photons, the *probability* of detecting a photon is $1/2M$ at all of Eve's ports except the one with the correct delay T_A. For that port, the probabilities are zero and $1/M$ respectively. The problem for Eve is that her photo-counts are almost completely randomly distributed: in effect, she has a serious signal-to-noise ratio problem.

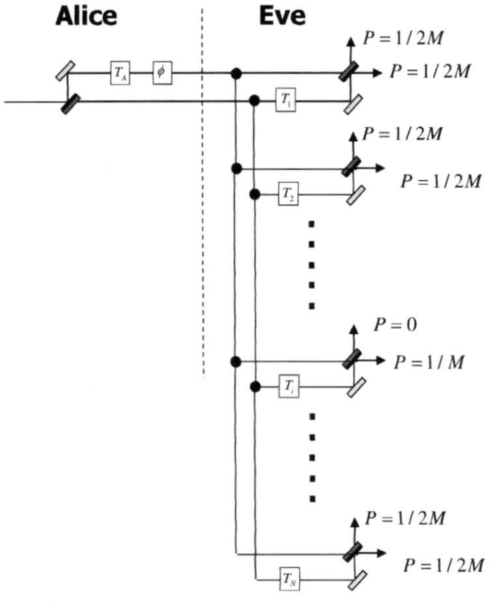

Figure 9. Eve's attack in photon counting case.

6. Example statistical analysis

We consider one specific example of an attack by Eve. Let the number of possible time delays M equal 10 and assume that in a given modulation interval Alice transmits a light pulse that results in 15 photo-counts. As illustrated in Figure 10, almost all of Bob's photocounts are at 0-bit output port (a single dark count appears at the 1-bit output). Eve's, on the other hand, are more or less randomly distributed. Bob can conclude with a high degree of certainty that a 0-bit was sent, whereas Eve's results are far from conclusive.

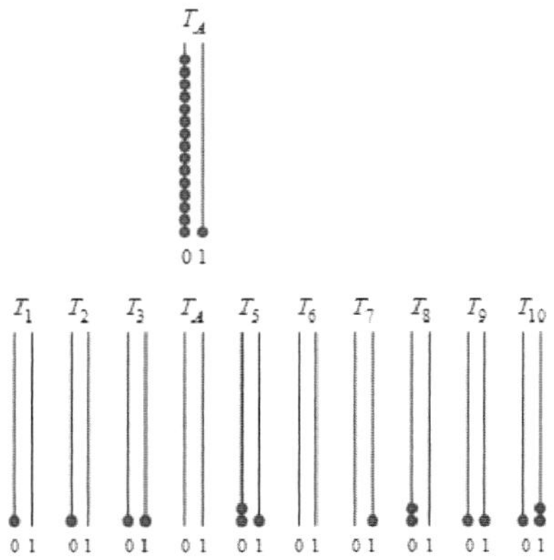

Figure 10. Photo counts for Bob and for Eve. Bob's clearly signify a 0-bit, whereas Eve's are essentially randomly distributed between 0-bits and 1-bits.

What can Eve do? In a simple attack, Eve can wait until she has ruled out all but a single time delay and associated phase value. She can then send light with the same characteristics on to Bob without betraying her presence. In order to do so, Eve must have access to a sufficiently large number of photons—a condition Alice and Bob will want to preclude. The detection of photons at interferometer output ports is governed by Poisson random processes. These processes are independent from output port to output port. The expected (mean) value of the photon count at a given output port is in proportion to the classical light intensity that would be observed at that port. Assuming that Eve has ideal components (consistent with the assumption that she has extraordinary capabilities), the probability that she determines the correct time delay T_A and associated bit value is given by $\left[1-\exp(-\bar{n}_{tot}/2M)\right]^{2(M-1)}\left[1-\exp(-\bar{n}_{tot}/M)\right]$ where $\bar{n}_{tot}$ is the mean number of detected photons and M is the number of different possible time delays.

The results of calculation of Eve's probability of identifying the correct time delay and phase for four different values (*M)* of possible coherence modulation time delays are shown in Figure 11. For the case of 10 possible time delays, Eve will require 65 photocounts on average (expected value) to achieve a probability exceeding 0.5 of selecting the correct time delay and bit value, information on both of which she must send to Bob in order to avoid detection of her presence. For a 0.9 probability of being correct, the mean number of photons she requires exceeds 100. Larger values of M reduce even further her probability of being correct.

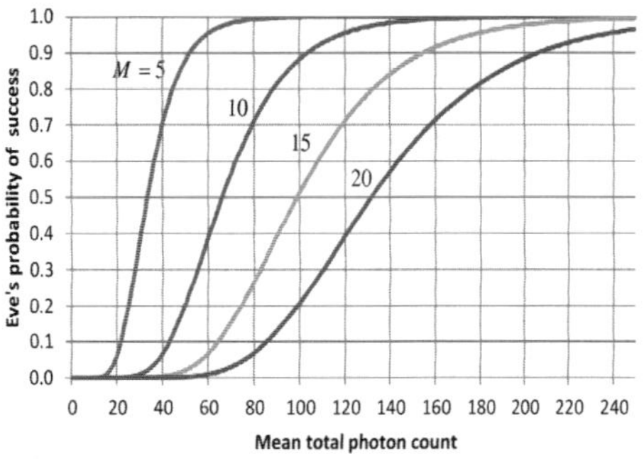

Figure 11. Eve's probability of determining correct time delay and bit value.

7. Concluding remarks

There are other attack strategies available to Eve, one of which is alluded to in Ref. [3]. All must be analyzed in detail. Various other points are noted here for consideration.

- Alice and Bob will want to operate in such a way that the mean number of photocounts available to Eve in a modulation interval is, with sufficiently high probability, less than the number she requires for a sufficiently high probability of correctly determining T_A and ϕ.
- The number of possible time delays can range from tens to perhaps thousands, depending on the technology used. Clearly the more the better.
- We assume that Eve can work with a perfect system—unity quantum detection efficiency, no dark count, etc.—whereas Bob and Alice face real-world limitations.
- The fraction of information Eve can gain with each transmitted bit is determined by the laws of physics. Quantum mechanics prevents Eve from cloning the photons to allow successful exhaustive parallel searches for the information.
- Whereas for maximal security quantum key distribution (QKD) requires that only a single photon be in the system at a time, the coherence modulation scheme allows—in fact encourages—multiple photons per bit, the mean number increasing with increasing numbers of possible time delays.
- The coherence modulation scheme allows for the transmission of general messages—plaintext, ciphertext, or keys for one-time pad use—with quantifiable security guaranteed by laws of physics.
- A minor modification of the system, to be described in another paper, allows direct implementation of the BB84 QKD protocol.
- Eve can betray her presence between Alice and Bob in a variety of ways. The various possible ways are currently under investigation.
- The effects of non-ideal components, dark current, quantum efficiency of detectors, etc., can be reduced through classical coding of the transmitted bit stream to compensate for a noisy channel.
- Difficulty of maintaining optical pathlength differences to fractions of wavelengths over long distances of, e.g., fiber, is a major concern. Candidate common-path interferometer implementations are under investigation.

References

[1] Barnett S M and Phoenix S J D 1996 *Philosophical Transactions: Mathematical, Physical and Engineering Sciences* **354** (1708) 793
[2] Gutierrez-Martinez C, Porte H, and Goedgebuer J P (1995) **10** 66
[3] Wacogne B and Jackson D A 1996 *IEEE Photon. Technol. Lett.* **8** 470

Analytical stabilization of modulated optical similaritons in a tapered graded-index waveguide

Amit Goyal[1,*], **Rama Gupta**[2], **Thokala Soloman Raju**[3] **and C N Kumar**[4]

[1] Department of Physics, GGDSD College, Chandigarh 160030, India
[2] Department of Physics, DAV University, Jalandhar 144008, India
[3] Indian Institute of Science Education and Research (IISER) Tirupati, Andhra Pradesh 517507, India
[4] Department of Physics, Panjab University, Chandigarh 160014, India

E-mail: *amit.goyal@ggdsd.ac.in

Abstract. We note that the self-similar waves recently reported, in a tapered graded-index waveguide, severely suffer from extreme increase in the amplitude, after traversing a finite distance. In this work, we theoretically described that by adding a suitable homogeneous background in the expression for widths of the self-similar solutions which is achievable through the modulation of tapering profile, we could avert this situation. We further find that the self-similar waves propagate in the nonlinear waveguide, quite a distance stably, albeit they undergo self-compression.

1. Introduction

In recent years, significant efforts have been posed in the study of design and properties of the tapered optical waveguides both theoretically as well as experimentally. Tapering is one of the best mechanisms for improving the coupling efficiency between the fibers and waveguides [1]. It produces high nonlinear effects even for low power intensity due to small mode area. One such effect is the enhancement of the two-photon absorption in the nonlinear tapered optical fibers in rubidium vapors [2]. Propagation of soliton and parabolic pulses through the tapered silica core optical fiber has been investigated numerically [3, 4]. It has significant effect on the wave dynamics and intensity of wave profile. Hence, a lot of work is being done on the propagation of the self-similar waves, waves which maintain their shape but the amplitude and width change with the modulating system parameters [5, 6], in the tapered graded index waveguides [7, 8, 9]. Tapering modulates the refractive index of the optical waveguide. For the tapered graded index optical waveguide, the refractive index will be of the form

$$n(x, z) = n_0 + n_1 F(z) x^2 + n_2 |u|^2, \tag{1}$$

where the first two terms correspond to the linear part of refractive index and the last term is Kerr-type nonlinearity. The function $F(z)$ describe the geometry of tapered waveguide along the waveguide axis. The quadratic variation of the refractive index in transverse direction is a good approximation to the true refractive index distribution, and is often known as lens-like medium. Such waveguides have applications in image-transmitting processes because they possess a very wide bandwidth. The shape of a taper i.e. $F(z)$ can be modeled appropriately depending upon the practical requirements.

In recent work, authors have studied the wave propagation through tapered graded index waveguides and reported the existence of self-similar waves [7, 8, 10]. The width of these waves obeys the second order differential equation identical to a wave equation governing the modes of an inhomogeneous planar waveguide with the refractive index profile given by the function $F(z)$. They have considered the lowest-order mode of sech^2-profile waveguides to study the propagation of self-similar waves. A close inspection reveals that for long distance propagation, these waves undergo catastrophic collapse. In this paper, we show that by modulating the tapering and width functions one can suppress the collapse of these self-similar waves. This has been achieved by adding a suitable homogeneous background in the expression of width function which can be experimentally achieved through the modulation of tapering profile. This analysis has been done for self-similar 1- and 2- soliton solutions, and first- and second-order rogue waves. Solitons [11] and rogue waves [12, 13] are the solutions of standard nonlinear Schrödinger equation which appear under different initial conditions. Modifying the width and tapering functions help us to propagate the self-similar solitons quite a long distance stably and also to control the exponential growth of background in self-similar rogue waves. Due to this modification in width function, the high amplitude feature of rogue waves no longer remains prominent.

2. Model equation and self-similar solutions

The propagation of beam through tapered graded-index nonlinear waveguide amplifier is governed by generalized nonlinear Schrödinger equation (GNLSE)

$$i\frac{\partial u}{\partial z}+\frac{1}{2k_0}\frac{\partial^2 u}{\partial x^2}+\frac{1}{2}k_0 n_1 F(z)x^2 u-\frac{i(g(z)-\alpha(z))}{2}u+k_0 n_2|u|^2 u=0, \tag{2}$$

where $u(x,z)$ is the complex envelope of the electrical field, g and α account for linear gain and loss, respectively, $k_0 = 2\pi n_0/\lambda$, λ being the wavelength of the optical source generating the beam, n_1 is the linear defocusing parameter ($n_1 > 0$), and n_2 represents Kerr-type nonlinearity. The dimensionless profile function $F(z)$ can be negative or positive, depending on whether the graded-index medium acts as a focusing or defocusing linear lens. Introducing the normalized variables $X = x/w_0$, $Z = z/L_D$, $G = [g(z) - \alpha(z)]L_D$, and $U = (k_0|n_2|L_D)^{1/2}u$, where $L_D = k_0 w_0^2$ is the diffraction length associated with the characteristic transverse scale $w_0 = (k_0^2 n_1)^{-1/4}$, equation (2) can be rewritten in a dimensionless form [7],

$$i\frac{\partial U}{\partial Z}+\frac{1}{2}\frac{\partial^2 U}{\partial X^2}+F(Z)\frac{X^2}{2}U-\frac{i}{2}G(Z)U+|U|^2U=0. \tag{3}$$

The similariton solutions for equation (3) can be obtained by transforming it into standard homogeneous NLSE using the gauge and similarity transformation [7, 14]

$$U(X,Z)=\frac{1}{W(Z)}\Psi\left[\chi(X,Z),\zeta(Z)\right]e^{i\Phi(X,Z)}, \tag{4}$$

where $W(Z)$ and X_c are the dimensionless width and position of the self-similar wave center. Substituting equation (4) into equation (3), one obtains the NLSE

$$i\frac{\partial \Psi}{\partial \zeta}+\frac{1}{2}\frac{\partial^2 \Psi}{\partial \chi^2}+|\Psi|^2\Psi=0. \tag{5}$$

Here the similarity variable, guiding-center position, effective propagation distance and phase are given respectively as [14], $\chi(X,Z) = \frac{X-X_c(Z)}{W(Z)}$, $X_c(Z) = W(Z)\left(\frac{X_0}{W_0}+C_{02}\int_0^Z \frac{dS}{W^2(S)}\right)$, $\zeta(Z) = \zeta_0+\int_0^Z \frac{dS}{W^2(S)}$, $\Phi(X,Z) = \frac{X^2}{2W}\frac{dW}{dZ}+\frac{C_{02}X}{W}-\frac{C_{02}^2}{2}\int_0^Z \frac{dS}{W^2(S)}$, where C_{02}, X_0 and W_0 are constant of integration. Further, the tapering function, gain and similariton width $W(Z)$ are related as

$$d^2W/dZ^2 - F(Z)W = 0, \tag{6}$$

and

$$G(Z) = -d[\ln W(Z)]/dZ. \tag{7}$$

As stated earlier, transformation given by equation (4) reduces equation (3) to the standard NLSE, so for all the localized solutions of NLSE, 1-, 2- soliton [11] and first- and second-order rational solutions [13], the corresponding optical self-similar solutions of equation (3) can be obtained by means of the reverse transformation variables and functions.

The general expression of the intensity for self-similar 1- and 2-soliton solutions (I_{1S} and I_{2S}, respectively) is given by

$$I_{1S}(X, Z) = \frac{\eta^2}{W^2}\text{sech}^2[\eta(\chi - v\zeta)], \tag{8}$$

where η and v are the soliton amplitude and velocity.

$$I_{2S}(X, Z) = \frac{16}{W^2}\frac{|N(\chi,\zeta)|^2}{D(\chi,\zeta)^2}, \tag{9}$$

where $N(\chi,\zeta) = \eta_1 \cosh\xi_2 \exp(-i\chi_1)[(\kappa_2-\kappa_1)^2+2i\eta_2(\kappa_2-\kappa_1)\tanh\xi_2+\eta_1^2-\eta_2^2]+\eta_2\cosh\xi_1\exp(-i\xi_2)[(\kappa_2-\kappa_1)^2-2i\eta_1(\kappa_2-\kappa_1)\tanh\xi_1-\eta_1^2+\eta_2^2]$, $D(\zeta,\chi) = \cosh(\xi_1+\xi_2)[(\kappa_2-\kappa_1)^2+(\eta_2-\eta_1)^2]+\cosh(\xi_1-\xi_2)[(\kappa_2-\kappa_1)^2+(\eta_2+\eta_1)^2]-4\eta_1\eta_2\cos(\chi_2-\chi_1)$. The arguments and the phases are given as $\xi_i(\chi,\zeta) = 2\eta_i\chi+4\eta_i\kappa_i\zeta$ and $\chi_i(\chi,\zeta) = 2\kappa_i\chi+2(\kappa_i^2-\eta_i^2)\zeta$, respectively, where, η_i and κ_i defines the amplitude and the velocity of the ith soliton ($i = 1, 2$) respectively.

The intensity expression for first- and second-order rational solutions (I_{R_1} and I_{R_2}, respectively) is given by [8]

$$I_{R_1}(X, Z) = \frac{1}{W^2}\left[1 + 8\frac{1+4\zeta^2-4\chi^2}{(1+4\zeta^2+4\chi^2)^2}\right], \tag{10}$$

$$I_{R_2}(X, Z) = \frac{1}{W^2}\left[\left(\frac{D-G}{D}\right)^2 + \left(\frac{H}{D}\right)^2\right], \tag{11}$$

where $D = \frac{1}{3}\left(\chi^2+\zeta^2\right)^3 + \frac{1}{4}\left(\chi^2-3\zeta^2\right)^2 + \frac{3}{64}\left(12\chi^2+44\zeta^2+1\right)$, $G = \left(\chi^2+\zeta^2+\frac{3}{4}\right)\left(\chi^2+5\zeta^2+\frac{3}{4}\right) - \frac{3}{4}$, $H = \zeta\left(\zeta^2-3\chi^2+2(\chi^2+\zeta^2)^2-\frac{15}{8}\right)$.
From equation (6), the width of similaritons obeys the second order differential equation identical to a wave equation governing the modes of an inhomogeneous planar waveguide with the refractive index profile given by the function $F(z)$. In earlier works, authors have considered the sech-type [7, 8, 14] and sech tanh-type [10] width profiles. The former in this case vanishes at infinity and the later has a node at $Z = 0$ which following equation (8) results in blow up of intensity. Here we show that adding a homogeneous background in the width profile, which in turn modifies the tapering profile of the waveguide, can help to avert this problem. We discuss these two cases elaborately below.

Case (a): By adding a constant background (α) into the width function, the modified width and tapering functions satisfying equation (6) can be written as

$$W(Z) = \alpha + \text{sech}(\beta Z) \tag{12}$$

and

$$F(Z) = \frac{\beta^2\,\text{sech}(\beta Z)\left(1-2\,\text{sech}^2(\beta Z)\right)}{\alpha+\text{sech}(\beta Z)}. \tag{13}$$

The gain can be found from equation (7) as

$$G(Z) = \frac{\beta\,\text{sech}(\beta Z)\tanh(\beta Z)}{\alpha+\text{sech}(\beta Z)}. \tag{14}$$

For $\alpha = 0$ and $\beta = 1$, these functional forms reduce to the one considered in Refs. [7, 8, 14]. The variation of the magnitude of the width, tapering and gain functions along the propagation direction, for different choices of α, is shown in figure 1.

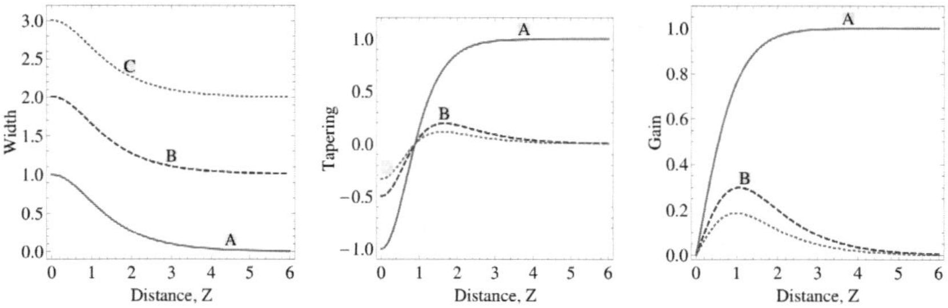

Figure 1. Profiles of width, tapering and gain given by equations (12) to (14) for different choices of α. Curve A (solid) for $\alpha = 0$; Curve B (dashed) for $\alpha = 1.01$; Curve C (dotted) for $\alpha = 2$. All the curves are plotted for $\beta = 1$.

It can be seen from the plots of width function that for $\alpha = 0$ width profile approaches zero for finite Z. But for non-zero values of α, width profiles approach to some finite value which will help us to control the collapse of self-similar waves. The addition of homogeneous background into the width profile also has a significant effect on the plots of tapering and gain functions. For non-zero values of α, both tapering and gain profiles approaches to zero unlike for the case $\alpha = 0$. It means that we need both tapering and gain medium only in small range of Z. It is also clear that tapering is initially negative and crosses zero at some value of Z, implying that the linear inhomogeneity of the waveguide is changing from focusing to defocusing type, but the magnitude of tapering is small for non-zero values of α which is desirable as it is easy to fabricate them, for practical applications.

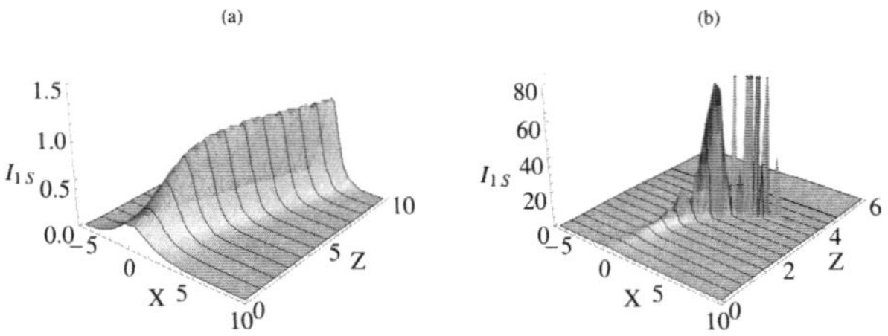

Figure 2. Comparison of intensity profiles of self-similar 1-soliton solution (equation (8)) for width function, $W(Z) = \alpha + \text{sech}(\beta Z)$ for (a) $\alpha = 1.01$ (Present work), (b) $\alpha = 0$ (Collapse). The parameters used in the plots are $\beta = 1, C_{02} = 0.3, X_0 = 0, \zeta_0 = 0, v = 0.3$ and $\eta = 1$.

In the following, we discuss the effect of modified width function on the intensity of self-similar waves for tapering given by equation (13). The intensity profile of self-similar 1- and 2-soliton solutions is shown in figures 2 and 3, respectively. Figures 2(a) (for $\alpha = 1.01$) and 2(b) ($\alpha = 0$), depicts the evolution of bright similariton, such as figure 2(a) depicts the self-compression of similariton for low values of z and it has constant intensity profile due to effect of modified tapering, where as figure 2(b) depicts the continuous self-compression of bright similariton and undergo catastrophic collapse. The similar comparison has also been made for evolution of self-similar 2-soliton solution, shown in figure 3. Hence, for both 1- and 2-soliton solutions, comparing the profiles for $\alpha = 1.01$ (figures 2(a) and 3(a)) with the profiles for $\alpha = 0$ (figures 2(b) and 3(b)), respectively, we find that the addition of homogeneous background in the width profile prevents the collapse of self-similar waves and allows them to propagate stably to a large distance. After simplifying the intensity expression it is easy to convince that $\alpha \in \mathbb{R} - [-1, 1]$, so as to have well behaved solutions. Choice of α poses inverse effect on intensity, hence its value should be adjusted as per physical requirement.

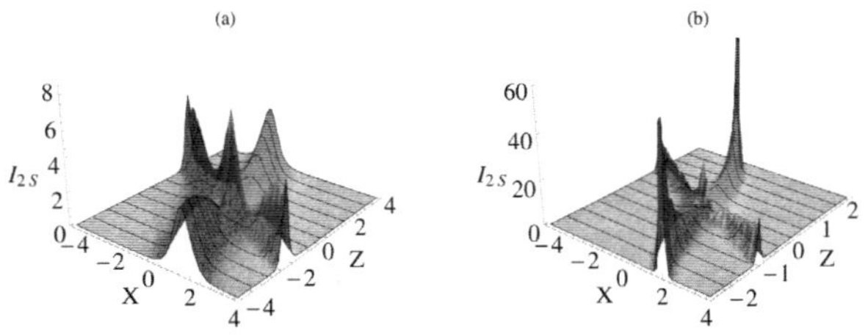

Figure 3. Comparison of intensity profiles of self-similar 2-soliton solution (equation (9)) for width function, $W(Z) = \alpha + \text{sech}(\beta Z)$ for (a) $\alpha = 1.01$ (Present work), (b) $\alpha = 0$ (Collapse). The parameters used in the plots are $\beta = 1, C_{02} = 0.3, X_0 = 0, \zeta_0 = 0, \eta_1 = 1, \eta_2 = 1.5, \kappa_1 = 0.5$ and $\kappa_2 = 2.5$.

In figures 4 and 5, we have shown the intensity profiles of self-similar first- and second-order rational solutions, respectively. These solutions are localized in both directions and have amplitude significantly larger as compared to the background. In general, these rational solutions are called rogue waves, but due to modification in width function, the high amplitude feature of rogue waves no longer remains prominent here, and we called them only rational solutions. In earlier work [8], it is found that these solutions are well behaved with $\alpha = 0$ for low range of z but for large propagation distance it shows exponential rise in the background. Here also the modified width function has significant effect on the intensity profiles of rational solutions as it prevents this exponential rise of background, as shown in figures 4(a) and 5(a), for $\alpha = 1.01$, unlike for the case with $\alpha = 0$ (figures 4(b) and 5(b)). Thus it makes the features of these waves more prominent.

Case (b): Corresponding to the choice of $W(Z) = \text{sech}Z \tanh Z$ [10], adding a homogeneous background to the width profile, the new width function can be written as

$$W(Z) = \alpha + \text{sech}(\beta Z)\text{tanh}(\beta Z) \tag{15}$$

and hence following equation (6), we obtain the following expression for tapering profile:

$$F(Z) = \frac{\beta^2 \,\text{sech}(\beta Z)\text{tanh}(\beta Z)\left(1 - 6\,\text{sech}^2(\beta Z)\right)}{\alpha + \text{sech}(\beta Z)\text{tanh}(\beta Z)}. \tag{16}$$

The gain function can be found from equation (7) as

$$G(Z) = \frac{\beta\, \text{sech}(\beta Z)\left(1 - 2\,\text{sech}^2(\beta Z)\right)}{\alpha + \text{sech}(\beta Z)\text{tanh}(\beta Z)}. \tag{17}$$

For $\alpha = 0$ and $\beta = 1$, these functional forms are considered in Ref. [10]. The variation of magnitude of the width, tapering and gain functions along the propagation direction, for different choices of α, is shown in figure 6. Here also addition of background in width profile has same effect on all the three functions as in the previous case. For non-zero values of α, width profiles approach to some finite value which is useful for the evolution of self-similar waves at large distances. The tapering and gain profiles approaches to zero at large propagation distances. It means waveguide should be tapered and doped with some gain medium only for finite range of z. Here, for non-zero values of α, tapering is always negative which implies that the linear inhomogeneity of the waveguide is only focusing. From figure 6 it is clear that non-zero values of α can be experimentally achieved by implementing the tapering and gain only in a small region of waveguide.

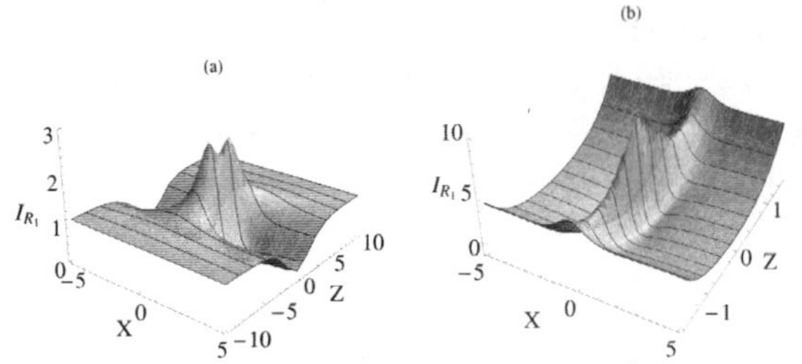

Figure 4. Comparison of intensity profiles of self-similar first-order rational solution (equation (10)) for width function, $W(Z) = \alpha + \text{sech}(\beta Z)$ for (a) $\alpha = 1.01$ (Present work), (b) $\alpha = 0$ (Collapse). The parameters used in the plots are $\beta = 1, C_{02} = 0.3, X_0 = 0$ and $\zeta_0 = 0$.

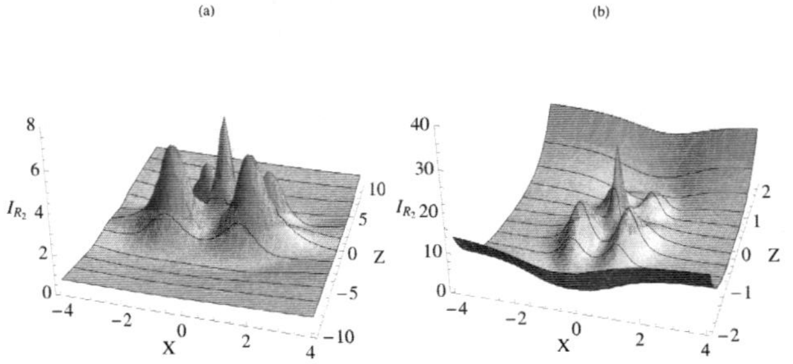

Figure 5. Comparison of intensity profiles of self-similar second-order rational solution (equation (11)) for width function, $W(Z) = \alpha + \text{sech}(\beta Z)$ for (a) $\alpha = 1.01$ (Present work), (b) $\alpha = 0$ (Collapse). The parameters used in the plots are $\beta = 1, C_{02} = 0.3, X_0 = 0$ and $\zeta_0 = 0$.

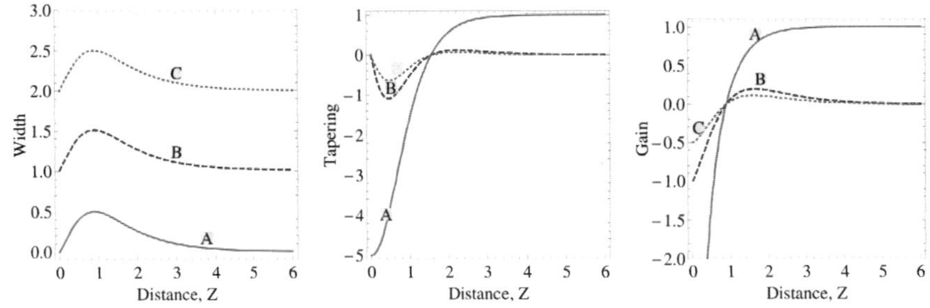

Figure 6. Profiles of width, tapering and gain given by equations (15) to (17) for different choices of α. Curve A (solid) for $\alpha = 0$; Curve B (dashed) for $\alpha = 1.01$; Curve C (dotted) for $\alpha = 2$. All the curves are plotted for $\beta = 1$.

As in the previous case, the intensity of self-similar waves can be found for tapering given by equation (16). In recent work [10], for $\alpha = 0$ and $\beta = 1$, a close inspection reveals that the intensity of bright similariton blow up at $z = 0$ and experiences a catastrophic collapse for large propagation distance. Due to the appearance of width function in the intensity expression of self-similar waves explicitly and also implicitly through χ and ζ, α plays an important role in the dominant appearance of wave over the background. By doing an exhaustive study of its available range, we find that wave appears more prominently for large values of z. Here we are specifically plotting the intensity profiles of self-similar

1- and 2-soliton, and first- and second-order rational solutions (figures 7 and 8, respectively), for $\alpha = 2$. Here too we observe that addition of a suitable homogeneous background in the width profile prevents the blow up of intensity and allows the wave to propagate to a large distance stably.

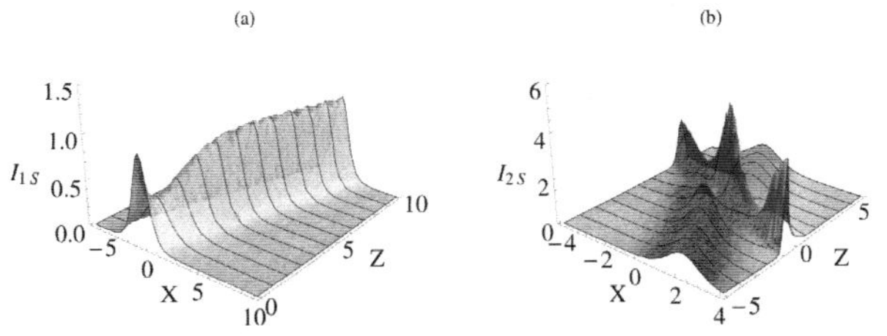

Figure 7. Intensity profiles of self-similar (a) 1-soliton and (b) 2-soliton solution for width function, $W(Z) = \alpha + \text{sech}(\beta Z)\tanh(\beta Z)$, for $\alpha = 2,\ \beta = 1$. The other parameters used are same as earlier.

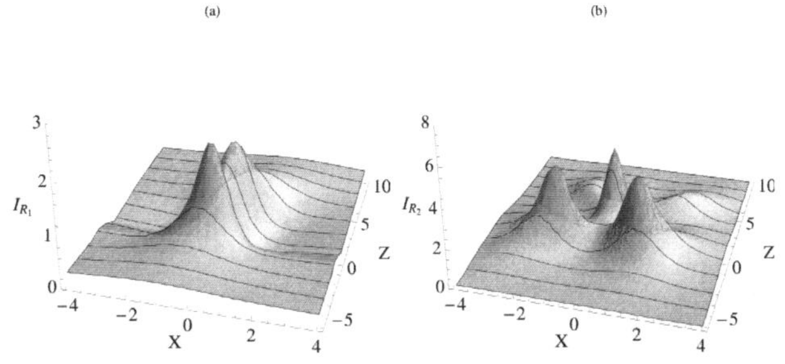

Figure 8. Intensity profiles of self-similar (a) first-order and (b) second-order rational solution for width function, $W(Z) = \alpha + \text{sech}(\beta Z)\tanh(\beta Z)$, for $\alpha = 2,\ \beta = 1$. The other parameters used are same as earlier.

3. Conclusion

We have noticed that the addition of a suitable homogeneous background in the expression for widths of the self-similar waves in a graded-index waveguide, through the modulation of tapering profile, enhances the stabilization of intensity profiles of the similaritons. Furthermore, we find that these self-similar waves propagate in the nonlinear waveguide, quite a distance stably, albeit they undergo self-compression. We hope that these similaritons can be launched in long-haul telecommunication networks, involving tapered graded-index waveguides.

Acknowledgement

T.S.R. wishes to thank DST, Government of India, for financial support through Young Scientist Award,during the course of this work.

References

[1] Almeida V R, Panepucci R R and Lipson M 2003 *Opt. Lett.* **28** 1302.
[2] Hendrickson S M, MLai M M, Pittman T B and Franson J D 2010 *Phys. Rev. Lett.* **105** 173602.
[3] Peacock A C 2010 *Opt. Lett.* **35** 3697.
[4] Peacock A C and Healy N 2010 *Opt. Lett.* **35** 1780.

[5] Kruglov V I, Méchin D and Harvey J D 2007 *J. Opt. Soc. Am. B* **24** 833.
[6] Dudley J M, Finot C, Richardson D J and Millot G 2007 *Nat. Phys.* **3** 597.
[7] Ponomarenko S A and Agrawal G P 2007 *Opt. Lett.* **32** 1659.
[8] Kumar C N, Gupta R, Goyal A, Loomba S, Raju T S and Panigrahi P K 2012 *Phys. Rev. A* **86** 025802.
[9] Dai C and Zhang J 2010 *Opt. Lett.* **35** 2651.
[10] Raju T S and Panigrahi P K 2011 *Phys. Rev. A* **84** 033807.
[11] Zakharov V E and Shabat A B 1972 *Sov. Phys. JETP* **62** 62.
[12] Peregrine D H 1983 *The J. Australian Math. Soc. B* **25** 16.
[13] Akhmediev N, Ankiewicz A and Taki M 2009 *Phys. Lett. A* **373** 675.
[14] Goyal A, Gupta R, Kumar C N, Raju T S and Panigrahi P K 2013 *Opt. Commun.* **300** 236.
[15] Dai C Q, Wang Y Y, Tian Q and Zhang J F 2012 *Ann. Phys.* **327** 512.

Tunneling dynamics of a two-atom system

G Muradyan, H Hakobyan and A Zh Muradyan

Laboratory for research and modeling of quantum phenomena, Yerevan State University, Yerevan 0025, Armenia

E-mail:gmurad@ysu.am

Abstract. We discuss the quantum tunneling of a diatomic molecule, concentrating on the transmission differences corresponding to different internal states. The evolution picture is constructed basing on introduction of notion of effective potential, the form of which is dependent on interatomic separation. To fully treat the tunneling problem analytically, we slightly modify the effective potential for each internal state. Clear splitting of these resonances enables us to suggest a scheme of molecular interferometer working as a Earth gravity sensor.

1. Introduction

Resonant tunneling is a basic phenomenon in elementary quantum mechanics [1] which appears when a particle tunnels through two (or more) potential barriers, and the incident energy coincides with any quasi-bound energy level inside the double- barrier. This is the matter-wave analog of what happens in an optical Fabry-Perot resonator. It was shown that for tunneling of particles with space-separated centers of mass [2], e.g. two-atom molecules, analogous phenomena is present even in case of bulk, non-split potential barriers. A number of aspects of resonant tunneling of structural particles have been studied theoretically (e.g. [3-6]), using mainly numerical simulations.

In this article we come to analytical expressions for the transmission and reflection amplitudes of symmetric and antisymmetric internal states of a diatomic molecule. On the scale of energies, the resonant transmission proves to be different for these states. As a conclusion, we suggest a scheme of a sensor using this difference to measure the Earth gravitation.

2. Resonant tunneling of symmetric and antisymmetric internal states

Let us consider the quantum tunneling of a molecule composed of two atoms with masses m_1, m_2 and separated by some distance l. Such 1D molecular structure may be modeled by a potential

$$V_{int} = -\frac{\alpha}{2}\left(\delta(x+l/2)+\delta(x-l/2)\right), \qquad (1)$$

where parameter α is the characteristic of molecule binding energy and $\delta(x)$ is the Dirac δ-function. In general we take the external (tunneling) potential for both particles in a rectangular form with different widths ($a_{1,2}$) and heights ($U_{1,2}$). Then the Hamiltonian takes the form

$$H = H_{rel} - \frac{\hbar^2}{2M}\frac{d^2}{dy^2} + U_1\theta\left(\left(a_1 - \frac{m_2 x}{m_1+m_2} - y\right)\left(a_1 + \frac{m_2 x}{m_1+m_2} + y\right)\right) + U_2\theta\left(\left(a_2 + \frac{m_1 x}{m_1+m_2} - y\right)\left(a_2 - \frac{m_1 x}{m_1+m_2} + y\right)\right), \quad (2)$$

where x and y denote the relative and the center of mass (c.m.) coordinates,

$$H_{rel} = -\frac{\hbar^2}{2m}\frac{d^2}{dx^2} + V_{int} \quad (3)$$

is the Hamiltonian of relative motion, m and M are reduced and total masses respectively. The eigenvalue and eigenfunction problem $H_{rel}\varphi_{s,a}(x) = -E_{s,a}\varphi_{s,a}(x)$ is a textbook one which has only two bound states, the *symmetric* one:

$$\varphi_s(x) = N_s\left(e^{\chi_s(x+l/2)}\theta(-x-l/2) + \frac{Cosh(\chi_s x)}{Cosh(\chi_s l/2)}\theta((x+l/2)(l/2-x)) + e^{-\chi_s(x-l/2)}\theta(x-l/2)\right), \quad (4)$$

with some energy value E_s determined from the transcendental equation

$$\sqrt{E_s} = \frac{\sqrt{2m}}{4\hbar}\alpha\left(1 - \exp\left(-\frac{\sqrt{2m}}{\hbar}l\sqrt{E_s}\right)\right), \quad (5)$$

and the *antisymmetric* one given by equations

$$\varphi_a(x) = N_a\left(e^{\chi_a(x+l/2)}\theta(-x-l/2) + \frac{Sinh(\chi_a x)}{Sinh(\chi_a l/2)}\theta((x+l/2)(l/2-x)) + e^{-\chi_a(x-l/2)}\theta(x-l/2)\right), \quad (6)$$

$$\sqrt{E_a} = \frac{\sqrt{2m}}{4\hbar}\alpha\left(1 - \exp\left(-\frac{\sqrt{2m}}{\hbar}l\sqrt{E_a}\right)\right), \quad (7)$$

Here $N_{s,a}$ are normalization constants and $\chi_{s,a} = \sqrt{2mE_{s,a}}/\hbar$. Note that the symmetric state exists for any parameters α and l, whereas the antisymmetric state occurs if $\alpha l > 4$.

In the following we will assume that the incident energy of a molecule is smaller than the absolute values of binding energies $E_{s,a}$. Then the continuous spectrum of internal states has no role in the transmission problem and may be omitted from the consideration. As a result,

$$\Psi(x,y) = \psi_s(y)\varphi_s(y) + \psi_a(y)\varphi_a(y), \quad (8)$$

where $\psi_s(y)$ and $\psi_a(y)$ are the wavefunctions of c.m. motion with respect to symmetric and antisymmetric internal states respectively. Starting from the full problem $H\Psi(x,y) = E\Psi(x,y)$, they obey equations

$$\left(-\frac{\hbar^2}{2M}\frac{d^2}{dy^2} + V_s(y) - (E+E_s)\right)\psi_s(y) + V_{sa}(y)\psi_a(y) = 0, \quad (9)$$

$$\left(-\frac{\hbar^2}{2M}\frac{d^2}{dy^2} + V_a(y) - (E+E_a)\right)\psi_a(y) + V_{as}(y)\psi_s(y) = 0, \quad (10)$$

with

$$V_{s,a}(y) = \int_{-\infty}^{\infty}\varphi_{s,a}{}^2(x)V(x,y)dx, \qquad V_{sa}(y) = V_{as}(y) = \int_{-\infty}^{\infty}\varphi_s(x)V(x,y)\varphi_{sa}(x)dx, \quad (11)$$

where $V(x,y)$ is the external potential in the Hamiltonian (2). The integrals in equations (11) are calculated and expressed through the long-length sum of elementary functions.

Homonuclear molecules. In case of same-type atoms $V_{sa}(y) = V_{as}(y) = 0$ and the translational wavefunctions satisfy the following separate equations of Schrödinger form:

$$\left(-\frac{\hbar^2}{2M}\frac{d^2}{dy^2}+V_s(y)\right)\psi_s(y)=(E+E_s)\psi_s(y), \tag{12}$$

$$\left(-\frac{\hbar^2}{2M}\frac{d^2}{dy^2}+V_a(y)\right)\psi_a(y)=(E+E_a)\psi_s(y). \tag{13}$$

Here $V_s(y)$ and $V_a(y)$ become exactly the "effective" potential energies corresponding to two internal states. Unfortunately, Eqs. (12) and (13) with such potential energies don't have analytic solutions. They will have them if the potential is modeled by a close "replica" form

$$V(y)=\begin{cases} \mathrm{v}-\mathrm{v}_0+\dfrac{\mathrm{v}_0}{\exp((-y+l_0)/a_0)+1} & 0\le |y|<y_0 \\ \dfrac{\mathrm{v}}{\exp((y-l_1)/a_1)+1} & |y|\ge y_0 \end{cases}. \tag{14}$$

Noting that $V_s(y)$ and $V_a(y)$ are symmetric functions of c.m. coordinate y, we proceed with symmetric and antisymmetric basic solutions of Eqs. (12), (13) denoting them as $\psi_{s,sym}(y)$, $\psi_{s,anti}(y)$ and $\psi_{a,sym}(y)$, $\psi_{a,anti}(y)$ respectively. This yields

$$\psi_s(y)=h_{s,sym}\,\psi_{s,sym}(y)+h_{s,anti}\,\psi_{s,anti}(y),\quad \psi_a(y)=h_{a,sym}\,\psi_{a,sym}(y)+h_{a,anti}\,\psi_{a,anti}(y). \tag{15}$$

So, in case of homonuclear molecules with two internal states, the problem of tunneling of the molecule splits into two separate tunneling problems for each internal state. Simultaneously, the problem of resonances transforms into the question whether the "effective" potentials in Eqs. (11) (or (14)) have internal minima.

Traditional left-side incident-wave geometry can be guaranteed through the well-disposed h-coefficients in Eqs. (15), with respect to asymptotic forms

$$\psi_{s,a}(y\to-\infty)=e^{ik_{s,a}y}+r_{s,a}e^{ik_{s,a}y},\quad \psi_{s,a}(y\to+\infty)=t_{s,a}e^{ik_{s,a}y}, \tag{16}$$

where $r_{s,a}$ and $t_{s,a}$ are the sought reflection and transmission probability amplitudes.

The main question about the form of "effective" potentials is directly linked to the distance l between the atoms, as is exemplified in figure 1. Additionally, the formation of completely isolated barriers takes place at $l\approx 2a$.

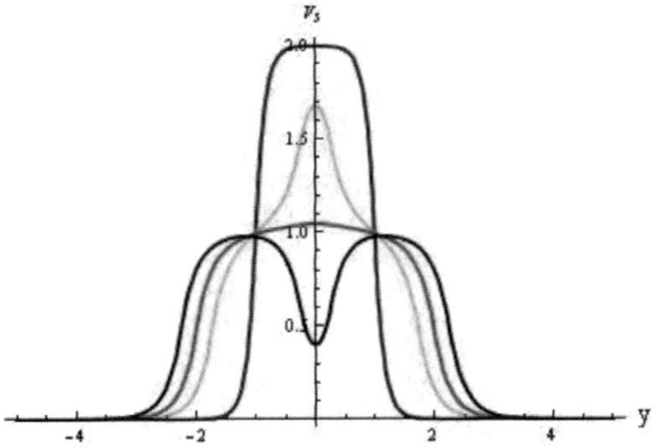

Figure 1. The effective potential $V_s(y)$ for different values of molecular length l for $\alpha/a E_R=4$. ($E_R=\hbar^2/2m a^2$). For blue line $l=0.1a$, for green $l=1.5a$, for red $l=2a$ and for black $l=2.5a$.

Analytic simulation expresses the transmission and reflection probability amplitudes, $r_{s,a}$ and $t_{s,a}$, through the product of exponential and Gauss hypergeometric functions. They are too cumbersome to be presented in the paper. To go ahead, we choose the length value $l=3.5a$, when the difference between effective potentials for the symmetric and antisymmetric internal states is still noticeable (figure 2).

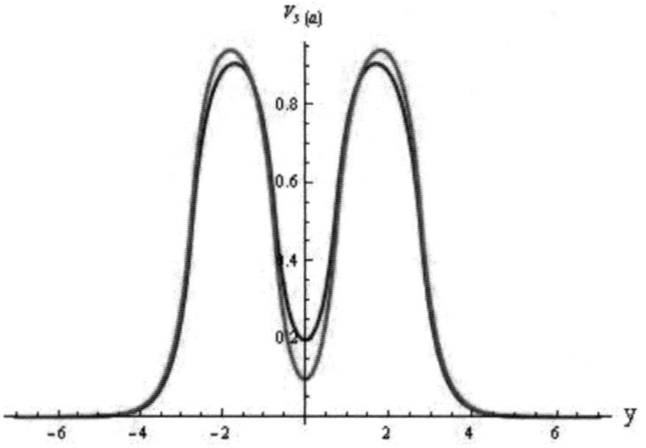

Figure 2. The effective potentials $V_s(y)$ (blue) and $V_a(y)$ (red) at $l = 3.5\,a$. The parameter of power of interatomic δ-function interaction ($\alpha / a E_R = 2.5$) is somewhat smaller than in figure 1.

It makes perceptible the difference between the corresponding quasi-bound energies and, as a consequence, forms a double-structure in the resonant tunneling picture, as it is depicted in figure 3. In fact, each of resonances works as a separator of symmetric and antisymmetric internal states, guiding them in opposite to barrier directions.

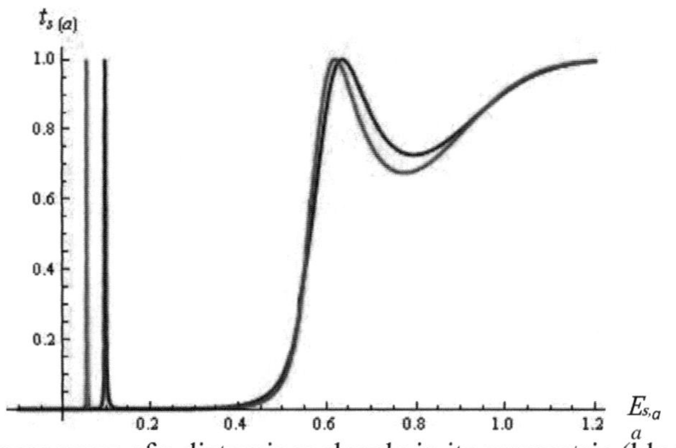

Figure 3. Quantum transparency of a diatomic molecule in its symmetric (blue) and antisymmetric (red) internal states versus the incident energy. Effective potentials are taken from figure 2. The resonances are posed at $E_s = 0.40\,E_R$ and $E_a = 0.38\,E_R$.

3. Conclusion. A scheme of molecular interferometer

The resonances in the quantum tunneling of a diatomic molecule may be seen as a consequence of constructive interference between the matter waves corresponding to constituent atoms being in different parts of the barrier. The mean interatomic distance must exceed the width of the barrier. This important restriction leaves out of interest the laser induced optical potentials. As a tunneling potential we see perspective in a single-layer atomic structures, such as the famous graphene. Our suggestion assumes a molecular Bose-Einstein condensate, loaded into a potential well over the graphene potential barrier. Switching off the trapping potential, the cold molecular cloud due to gravity falls on the graphene layer having resonant transmission energy for the antisymmetric internal state E_a. The antisymmetric state will tunnel through the layer while the symmetric state will bounce upwards. Immediately after that, the graphene layer will be shifted down to some appropriate position to match the energy $E_s > E_a$. As a result, the retardation of the symmetric state will be a linear function of the gravitation acceleration down the layer.

Figures 2 and 3, aside the illustration purpose, roughly represent a situation with cold Na_2 molecules [7, 8], where the role of internal states is played by the rotation spectrum. Corresponding resonant transmission energies are about $0.4 \cdot 10^{-4}\, eV$, giving *1.2 m* height for the position of the magnetic trap above the graphene layer.

It has to be noted that atomic and molecular interferometers, in parallel to pendulum gravimeters, have repeatedly been used as local gravity sensors (see, for instance, [9-14]). The present scheme of measurement represents a new type on this road: it substitutes material or light-pulse gratings by a single graphene layer. Another principle advantage of the scheme is the irrelevance of complete resonant transmission relative to specific values of parameters of the barrier potential.

Acknowledgement
This work was supported by funds from State Committee of Science of Armenia and Armenian Science and Education Fund grant nano-3488.

References
[1] Penley J C 1962 *Phys. Rev.* **128** 596
[2] Saito N and Kayanuma Y 1994 *J. Phys.: Condens. Matter* **6** 3759
[3] Pen'kov F M 2000 *Phys. Rev.* A **62** 044701
[4] Shegelski M R A *et al* 2008 *Phys. Rev.* A **77** 032702
[5] Gusev A A *et al* 2014 *Phys. of Atomic Nuclei* 77 389
[6] Muradyan G and Muradyan A Zh 2015 *Few Body Physics (submitted)*
[7] Chapman M S *et al* 1995 *Phys. Rev. Lett.* **74** 4783
[8] Xu K *et al* 2003 *Phys. Rev. Lett.* **91** 21040
[9] Kasevich M and Chu S 1991 *Phys. Rev. Lett.* **67** 181
[10] Snadden M J 1998 *Phys. Rev. Lett.* **81** 971
[11] Ott H *et al* 2005 *Physics* **15** 82
[12] Rosi G *et al* 2014 *Nature* **510** 518
[13] Hamilton P 2015 *Phys. Rev. Lett.* **114** 100405
[14] Hartwig J 2015 *New J. Phys.* **17** 035011

Investigation of spectroscopic properties of $LiNbO_3:Ho^{3+}$ crystals

Hasmik G Demirkhanyan

Department of Theoretical Physics and Physics Learning Methods, Armenian State Pedagogical University after Kh. Abovyan, 375010, Yerevan, Armenia

E-mail demirkhanyan_hasmik@yahoo.com,

Abstract. In this paper the Stark problem for Ho^{3+} ion ($4f^{10}$ electroni c configuration) in $LiNbO_3$ crystal is solved. Main spectroscopic parameters induced by inter-Stark transitions are determined. Based on analysis of experimental and theoretical data of the main spectroscopic characteristics of $LiNbO_3$:Ho3+ crystals, it's perceptiveness as a material for optical cooling devices is shown.

1. Introduction

Lithium Niobate crystals $LiNbO_3$ (LN), doped by Rare Earth (RE^{3+}) are perspective materials for compact multifunction lasers in the infrared and visible regions of the spectrum[1,2], as well as for optical cooling systems and self-cooled lasers [3,4]. Optical spectra of impurity absorption and emission in LN-Ho^{3+}crystal are investigated in [5]. Particularly, the energetic scheme of Stark levels of $^5I_8$, $^5I_7$, $^5I_6$ and $^5I_5$ multiplets, standard Judd-Ofelt analysis is provided. At the same time, optical spectra of LN-Ho^{3+}crystal in 1.8 -2.2 μm wavelength range, caused by inter-Stark transitions of ground ($^5I_8$) and first excited ($^5I_7$) multiplets are little studied experimentally. Meanwhile, these transitions are interesting for LN-Ho^{3+}optical cooling systems. Theoretical investigation of spectroscopic properties of LN-Ho^{3+} crystal in1.8 -2.2 μm wavelength range is given in [6].

In this work spectroscopic properties of LN-Ho^{3+} crystal in 1.5-2.0μm wavelength range in order to identify the possibility of establishing optical cooling systems and self-cooled lasers based on LN-Ho^{3+} crystal are discussed.

2. Spectroscopic characteristics of LN-Ho^{3+} crystal

It is well known, that in the crystal field (CF) of LN (point symmetry C_{3v}) ground ($^5I_8$) and first excited ($^5I_7$) multiplets of Ho^{3+} ion are splitting to 11th and 10th Stark sublevels respectively [6], transitions between which are defined the spectroscopic properties of the LN-Ho^{3+}crystal.

As known, the intensive lines of impurity absorption and emission spectra of RE^{3+} doped crystals are caused by indirect electro-dipole (IED) and magneto-dipole (MD) transitions, thus the line straight of transition is determined by expression $S_{i\to f} = \chi_{ed} \cdot S_{i\to f}^{(ied)} + \chi_{md} \cdot S_{i\to f}^{(md)}$ (where $S_{i\to f}^{(ied)}$ and $S_{i\to f}^{(md)}$ are line straight of IED and MD transitions, χ is local field correction). Un-polarized line straights of inter-stark transitions are [6]: $S_{i,f}^{(ied)} = \sum_t \Omega_t A_t(i,f)\left|\langle J_f \| U_t \| J_i\rangle\right|^2$ and $S_{i,f}^{(md)} = A_{md}(i,f) S_{md}$. Here $\Omega_2 = 4.3\times10^{-20}$, $\Omega_4 = 5.11\times10^{-20}$, $\Omega_6 = 1.89\times10^{-20}$ cm^2 are Judd - Ofelt's parameters [5],

$\langle {}^5I_7\|U_2\|\,{}^5I_8\rangle = -0.1477$, $\langle {}^5I_7\|U_4\|\,{}^5I_8\rangle = 0.3338$, $\langle {}^5I_7\|U_6\|\,{}^5I_8\rangle = -1.2852$ are reduced matrix elements of the irreducible unit operator U_t, $S^{(md)} = 51\mu^2/e^2$ is line straight of multiplet-to-multiplet MD transition (μ is Bohr's magneton) [6]. Coefficients

$$A_t^{(ied)}(i,f) = \sum_{m=-t}^{t}\left|\sum_{M_i,M_f}(-1)^{J_f-M_f}\, a_{J_fM_f}^{*(f)} a_{J_iM_i}^{(i)} \begin{pmatrix} J_f & t & J_i \\ -M_f & m & M_i \end{pmatrix}\right|^2 \quad (1)$$

$$A_{md}(i,f) = \sum_{m=-1}^{1}\left|\sum_{M_i,M_f}(-1)^{J_f-M_f}\, a_{J_fM_f}^{*(f)} a_{J_iM_i}^{(i)} \begin{pmatrix} J_f & 1 & J_i \\ -M_f & m & M_i \end{pmatrix}\right|^2 \quad (2)$$

determine distribution of intensities of IED and MD transitions by Stark states. Here J is the total angular momentum, M is its projection, $a_{J_iM_i}^{(i)}$ are numerical coefficients in the wave functions of νth Stark state [6], $\begin{pmatrix} \cdot & \cdot & \cdot \\ \cdot & \cdot & \cdot \end{pmatrix}$ is 3j-symbols. Values of calculated line straights of actual transitions and main spectroscopic characteristics, such as absorption coefficients (α), probabilities of spontaneous transitions (A), luminescence branching ratio (β), absorption efficiencies (α) are given in tables 1 and 2.

Table 1. Parameters of ASF spectra of $LiNbO_3:Ho^{3+}$ at T = 300K.

transition	λ_{lum}, nm	$S\times10^{-21}$ cm^2	A, s^{-1}	$\delta = N_i/N_0$,	β , %
$\mu_3 \to \nu_5$	2020	4.78368	41.8486	0.1244	1.5
$\mu_4 \to \nu_5$	2013	4.3078	38.08	0.1146	1.3
$\mu_5 \to \nu_5$	2008	4.93298	43.933	0.1082	1.4
$\mu_6 \to \nu_6$	2021	7.64323	66.7652	0.1026	2.0

Table 2. Parameters of absorption spectra of $LiNbO_3:Ho^{3+}$ at T = 300K.

transition	λ_p,nm	S,10^{-21} cm^2	α, 10^{-26}cm^3	N_i/N_0,	η_{abs},%
$\nu_5 \to \mu_1$	2035	3.6653	1.4216	0.0754	0.8
$\nu_6 \to \mu_1$	2053	5.1332	2.0086		1.0
$\to \mu_2$	2042	6.5145	2.5354	0.0617	1.2
$\to \mu_3$	2037	4.0845	1.5858		0.8

3. $LiNbO_3:Ho^{3+}$ crystal as perspective optical cold material

It is known that for determining the spectral region of expected optical cooling and evaluation of the efficiency of this effect by means of the anti-Stokes fluorescence (ASF) method it is useful to use the following expression [7]

$$\gamma = \eta\left[1 - e^{-(\alpha_p+\alpha_b)L}\right]\frac{\lambda_p - \langle\lambda_f\rangle}{\langle\lambda_f\rangle}. \quad (3)$$

Here γ is the efficiency of the optical cooling effect, η is the quantum yield of fluorescence from the excited level, α_p is the absorption coefficient at the optical excitation wavelength λ_p, α_b is the background absorption of the matrix, L is the thickness of the sample along optical excitation, and $\langle\lambda_f\rangle$ is the mean wavelength of fluorescence: $\langle\lambda_f\rangle = \int_0^\infty \lambda\, I_f(\lambda)\, d\lambda / \int_0^\infty I_f(\lambda)\, d\lambda$, ($I_f$ is the spectral intensity of fluorescence). By using values of spectroscopic characteristics calculated in [6], it is easy to obtain $\langle\lambda_f\rangle = 2002\, nm$. In figure 1 the absorption spectra of LN - Ho^{3+} at 1.8 - 2.2 μ vawelength [5], the possible implementation of the optical cooling with CW pumping at wavelengths 2035–2053 nm are shown as well. Thus, the effect of optical cooling, on the basis of crystal LN -Ho^{3+}, can be obtained by ASF method at wavelengths λ > 2002nm. For estimation of efficiency of the optical cooling effect we assume that, first off, intensity of pumping are so far from saturation, second, $\alpha_b \ll \alpha_p$, and third, the probability of zero-phonon transition from $^5I_7$ multiplet is negligibly small, so the quantum yield from the excited Stark sublevel is mainly determined by the luminescence branching ratio. Then, assuming L=1cm and N $\approx 10^{21}cm^{-3}$, according to Tables 1 and 2, it is easy to estimate the value of γ for different schemes of pumping and emission wavelengths. Where in value of absorption coefficient $\alpha_p(\lambda)$ at λ wavelength is connected with value of α (Table 2) by expression $\alpha_p(\lambda) = \alpha(\lambda)N_i/\Delta\lambda$, where $N_i = \delta N$ is population of $v_i{}^{th}$ level (δ is Boltzmann factor). In numerical estimations we use the following values: $\Delta\lambda \approx 10nm$, N$\approx 10^{21}cm^{-3}$. Results of evaluation are given in Table 3. In Table 3 is given expected values of temperature decrease $\Delta T = \gamma\left(\varepsilon_{em} - \varepsilon_p\right)/k$,($\varepsilon_{em}$ is energy of anti-Stocks luminescence, $\varepsilon_p$ispump energy) as well. It is seen, that temperature of cooling of separate channel is varies in the range of 0.8 - 2.6 K and efficiency is 0.014 - 0.032%, which is comparable with the corresponding characteristics of other materials [8]. Since channels are independent to each other, the total efficiency will be: for 2035nm γ = 0.154% ($\Delta T = 10.83K$) and for 2050 nm γ = 0.17% ($\Delta T = 17.87K$).

Table 3. Temperature and efficiency of optical cooling at T = 300K.

transition	$\nu_5 \rightarrow \mu_1$		$\nu_6 \rightarrow \mu_1$		$\nu_6 \rightarrow \mu_2$		$\nu_6 \rightarrow \mu_3$	
	ΔT, K	γ, %	ΔT, K	γ, %	ΔT, K	γ, %	ΔT, K	γ, %
$\mu_3 \rightarrow \nu_5$	0.83	0.016	2.02	0.018	1.75	0.023	1.27	0.021
$\mu_4 \rightarrow \nu_5$	1.07	0.014	2.05	0.015	2.11	0.021	1.53	0.018
$\mu_5 \rightarrow \nu_5$	1.40	0.015	2.46	0.016	2.6	0.022	2.04	0.020
$\mu_6 \rightarrow \nu_6$	1.08	0.022	2.52	0.023	2.35	0.032	1.61	0.028

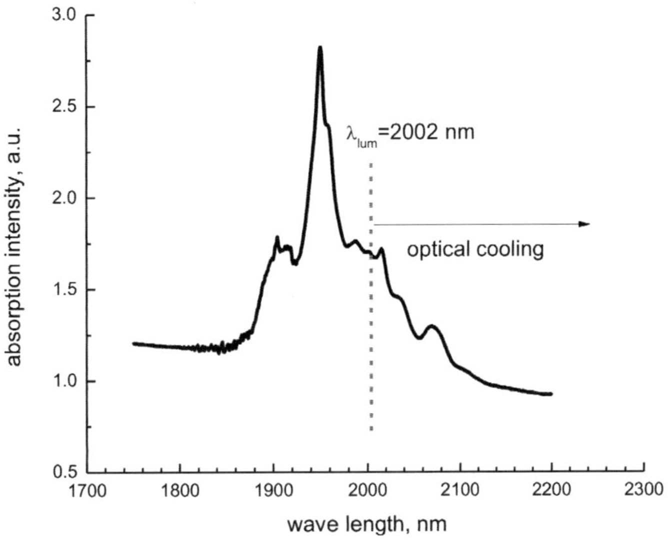

Figure 1. Absorption spectra of LN-Ho^{3+} at 1.8-2.2 μ [5]

4. Conclusion

Estimations show that the LN - Ho^{3+} crystal is perspective material for optical cooling systems. At the same time, it should be noted, that these estimations are rough enough. To get accurate values of efficiency of cooling, it is necessary to do more detailed experimental and theoretical investigations of spectroscopic characteristics of LN - Ho^{3+} crystal including the widths and shifts of spectral lines, as well as examination of their temperature dependences.

References

[1] Johnson L F and Ballman A A 1969 *J. Appl. Phys.* **40** 297
[2] Cordova-Plaza A, Digonnet M and Shaw H J 1987 *IEEE J. Quantum Electron.* QE-13262
[3] Demirkhanyan G G, Kokanyan E P, Gruber J B, Sardar D K and Pokhrel M 2014 *Proc. Conf. dedicated to the 90th anniversary of Armenian State Pedagogical University* pp. 383-388 (in russian)
[4] Babajanyan V G 2013 *Laser Phys.* **23** 126002
[5] Lorenzo A, Bausa L E, Sanz Garcia J A and Sole′ J G 1996 *J. Phys.: Cond. Matter* **8** 5781
[6] Demirkhanyan G G, Kokanyan E P and Demirkhanyan H G 2015 *J. of Cont. Phys. of Armenian Academy of Sciences* **50**(3) 318-329 (in russian)
[7] Melgaard S, Seletskiy D, Di Lieto A, Tonelli M and Sheik-Bahae M 2013 *Opt. Lett.* **38** 1588
[8] Nemova G and Kashyap R 2010 *Rep. Prog. Phys.* **73** 086501

Spectral domain soliton-effect self-compression

A P Grigoryan, G L Yesayan, A A Kutuzyan and L Kh Mouradian

Ultrafast Optics Laboratory, Faculty of Physics, Yerevan State University, Armenia

E-mail: akutuzyan@ysu.am

Abstract. On the basis of numerical studies, we demonstrate the nonlinear process of spectral self-compression in a fiber with anomalous dispersion, as a spectral analogue of soliton-effect self-compression (or soliton self-compression).

1. Introduction

The nonlinear process of spectral compression (SC) in a dispersive delay line (DDL) followed by a nonlinear fiber demonstrates promising applications to the signal analysis-synthesis problems in ultrafast optics [1]. In the SC system, the negative phase of pulse obtained in a DDL is compensating by the positive phase acquired by nonlinear self-phase modulation (SPM) in a fiber with normal dispersion [2, 3]. The impact of group velocity dispersion (GVD) on SC for sub-picosecond pulses in the range of normal dispersion, i.e. at wavelengths <1.3μm for standard silica fibers, is analyzed in [4] in view of shaping flattop pulses. In range of anomalous dispersion, i.e. at wavelengths >1.3μm for silica fibers, the combined impact of GVD and SPM leads to the formation of solitons [5, 6], when the contributions of GVD and SPM balance each other. The pulse self-compression phenomenon arises when the impact of SPM exceeds the GVD, and high-order solitons are shaped [7]. Recently, generation of sub-two-cycle pulses by soliton self-compression of 100-fs pulses from a Ti:sapphire laser at 85 MHz using a 4.85-mm long nonlinear photonic crystal fiber are demonstrated numerically [8] and experimentally [9]. The efficient soliton self-compression down to 5.07 fs is demonstrated also in As_2Se_3 and As_2S_3 chalcogenide photonic nanowires [10].

When the contribution of GVD exceeds the SPM impact in the fibers with anomalous dispersion, the spectral self-compression (self-SC), a spectral analogue of soliton self-compression, is anticipated. Recently, the soliton self-SC realization in a standard silica fiber with negative dispersion at the wavelength ≥ 1.3μm was predicted numerically [11] and observed experimentally [12]. The objective of our work is detailed numerical study of the soliton self-SC process revealing its nature, peculiarities and general regulations on the basis of physical pattern of the process.

2. Numerical studies

In our numerical experiments, we study the pulse and spectrum evolution during the propagation through a fiber with Kerr-nonlinearity and anomalous dispersion. We carry out simulations for initial Gaussian, secant-hyperbolic and super-Gaussian pulses.

The pulse propagation in a single-mode fiber is described by nonlinear Schrödinger equation [1]:

$$i\frac{\partial \psi}{\partial \zeta} = \frac{1}{2}\frac{\partial^2 \psi}{\partial \eta^2} + R|\psi|^2 \psi \quad (1)$$

where $\zeta = z / L_D$ is the dimensionless propagation distance normalized to the dispersive length $L_D = \tau_o^2 / |k_2|$ (k_2 is the coefficient of second-order dispersion); $\eta = (t - z/u)/\tau_0$ is the running time normalized to the initial pulse duration τ_0. The nonlinearity parameter R is given as $R = L_D / L_{NL}$, where $L_{NL} = (k_0 n_2 I_0)^{-1}$ is the nonlinearity length, n_2 is the Kerr index of silica, and I_0 is the peak intensity. The first and second terms of the right part of Eq. (1) describe the impact of GVD and nonlinearity, correspondingly. We use the split-step Fourier method for numerical solution of equation, with the fast Fourier transform algorithm on the dispersive step [13, 14].

We study soliton self-SC, which occurs when the dispersive length in the fiber is shorter than the nonlinear length ($L_D < L_{NL}$, i.e. $R < 1$). Therefore, at first the GVD stretches the pulse by acquiring a chirp. Afterwards, the accumulated impact of nonlinear SPM leads to the chirp compensation, and as a result, the spectrum becomes compressed. We study behavior of the pulse in a fiber with negative dispersion for different values of the nonlinearity parameter and fiber length.

The propagation mapping for Gaussian pulse and its spectrum for R = 0.6 are shown in Figures 1 (a, b), correspondingly.

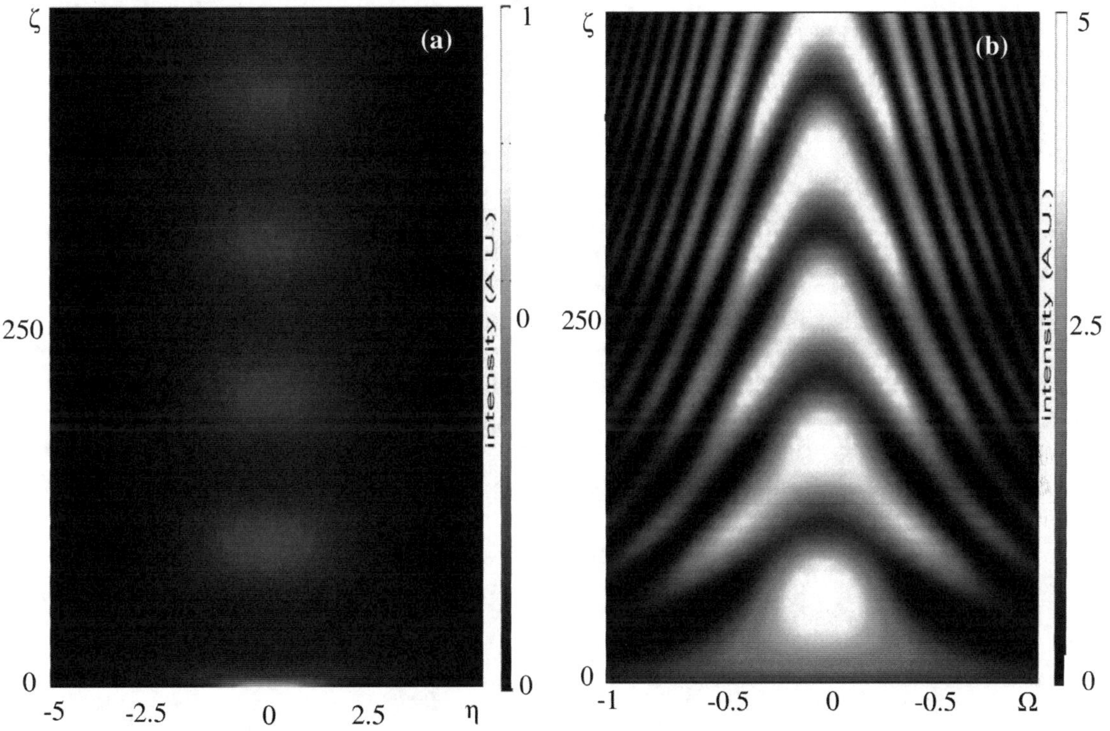

Figure 1. The 3D map of pulse self-interaction for pulse (a) and spectrum (b) (R=0.6, initially transform-limited Gaussian pulse).

In initial propagation step the pulse is stretched and spectrum is compressed. Afterwards, the width of central peak of spectrum decreases and the main part of the pulse energy transfers to spectral satellites. At a certain fiber length the reverse process starts – the pulse self-compresses. The process can be explained in the following way: in the initial propagation stage the spectrum is compressed, which leads to decreasing of the GVD impact. As a result, the dispersive length increases, therefore, the nonlinearity parameter increases. If the condition $R > 1$ is satisfied ($L_D > L_{NL}$), the pulse is compressed. Then, the spectrum stretches, which leads to the increase of GVD impact (decrease of L_D and *R*). When the condition $R<1$ ($L_D < L_{NL}$) is satisfied, the spectrum is compressed.

The above described process is of periodic character. However, for every next cycle the quality of the SC worsens as compared with the previous SC. In Figure 2, the peak value of spectrum and pulse is shown for initial Gaussian pulse. Our study clearly shows the periodic nature of the process for initial secant-hyperbolic pulse as well.

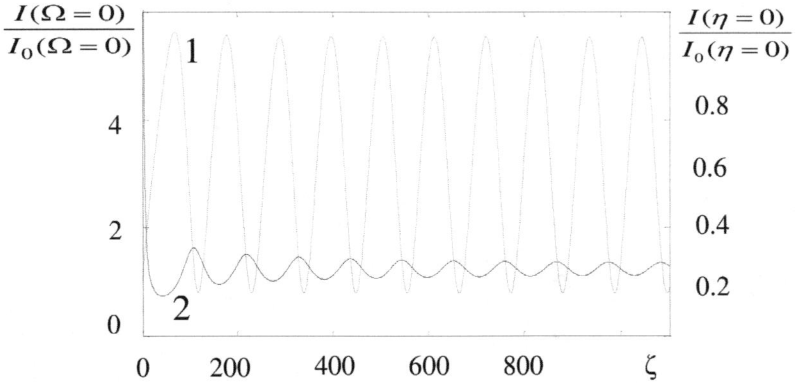

Figure 2. The peak value of spectrum (1) and pulse (2) vs fiber length for initial Gaussian pulse.

The behavior of spectrum in the propagation process is similar to the pulse behavior in the case of the soliton compression. First, the pulse is compressed on a distance equal to the period, and then it stretches. In our case, spectrum has similar behavior. However, because of incomplete chirp cancellation, the change in the spectrum does not exhibit strict periodicity. The study shows decrease of the process period with the reduction of nonlinearity parameter. For R=0.25, the SC ratio is ~ 4, without practical change of spectrum in the range of $\zeta \approx 11000\text{-}17000$ (Figure 3): the pulse large stretching (~6000^{X}) decrease the peak intensity of pulse which results in practical absence of the nonlinear self-interaction.

Figure 4 shows initial (dashed line) and compressed (solid line) spectra for initial transform-limited Gaussian ($A(t) = \exp(0.5 \cdot t/\tau_0)^2$) (a), secant-hyperbolic ($A(t) = \mathrm{sech}(t/\tau_0)$) (b), and super-G

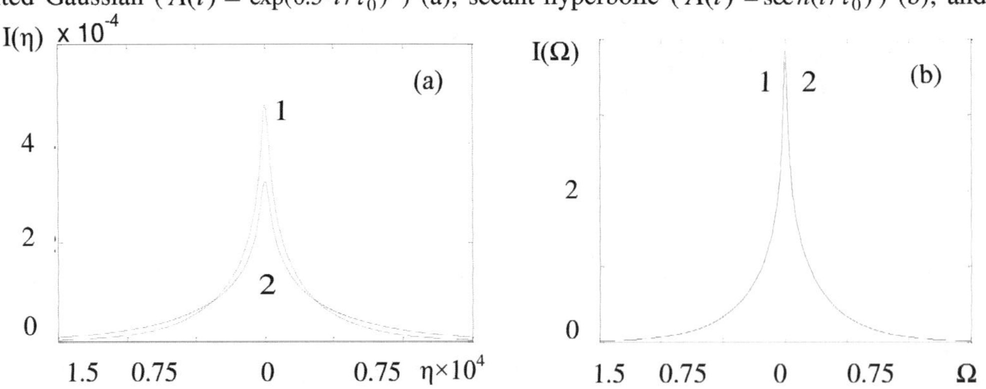

Figure 3. Pulses (a) and spectra (b) for $\zeta \approx 11000$ (1), $\zeta \approx 17000$ (2)

Figures illustrate the case of maximal soliton self-SC. The SC ratio ($\Delta\Omega_0/\Delta\Omega$) is ~ 6 for Gaussian pulse, obtained when the fiber length and nonlinearity parameter were $\zeta \approx 50$ and R = 0.6, correspondingly. For initial secant-hyperbolic and super-Gaussian pulses, the SC ratio is $\Delta\Omega_0/\Delta\Omega \approx 5$ ($\zeta \approx 81$, R = 0.4) and $\Delta\Omega_0/\Delta\Omega \approx 11$ ($\zeta \approx 80$, R= 0.7), correspondingly.

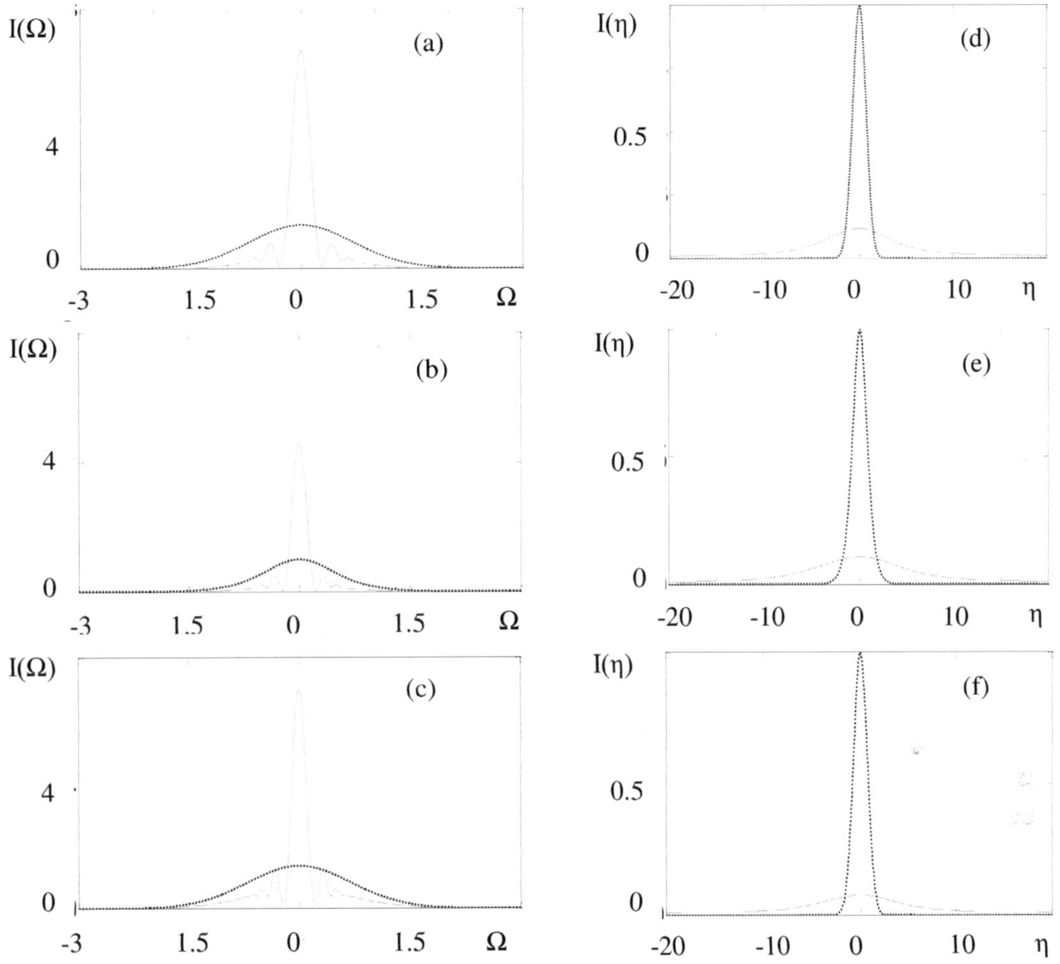

Figure 4. Initial (dashed line) and compressed (solid line) spectra for the initial transform-limited Gaussian (a), secant-hyperbolic (b) andsuper-Gaussian (c) pulses, which correspond to the case of maximal soliton self-SC, and pulses (d), (e), (f)), correspondingly.

The corresponding pulses are shown in Figure 4 (d, e, f). The coefficient of soliton self-SC quality K (defined as the ratio of energy in the central part of pulse to the whole energy) equals to ~ 0.8 for Gaussian, secant-hyperbolic and super-Gaussian pulses.

We have studied also the self-SC for slightly chirped pulses. The studies are carried out for the Gaussian pulses dispersively stretched (1.2^X, 1.4^X, and 2.2^X) and chirped (with the chirp coefficients C=0.83; 0.7 and 0.45) in the +/- dispersive medium. The results of studies for C= -0.7 are shown in Figure 5. In this case the SC factor is ~ 24, for R = 0.5, $\zeta \approx 700$, and $K \approx 0.3$. The same SC factor we have for the C=+0.7 and also for transform limited pulses with the parameters for R = 0.46 (adequate to the 1.4^X stretched pulse with R = 0.5), and $\zeta \approx 600$. This compassion shows that only the pulse peak intensity and spectral bandwidth are important for the process, and initial phase of pulses (in the range of C=+/- 0.45-0.83) does not impact to the process.

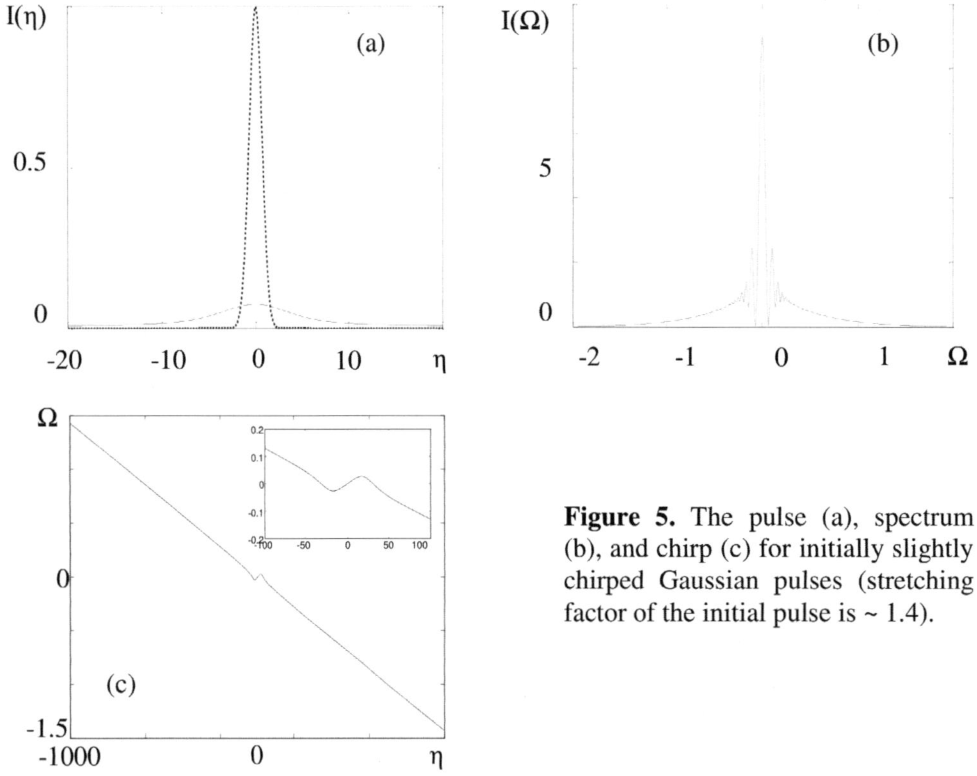

Figure 5. The pulse (a), spectrum (b), and chirp (c) for initially slightly chirped Gaussian pulses (stretching factor of the initial pulse is ~ 1.4).

3. Conclusion

Our detailed study have shown the soliton spectral self-compression in the fiber "directly", without dispersive delay line, in the fiber with anomalous dispersion. We have demonstrated that there is an analogy between the processes of soliton self-compression and soliton spectral self-compression. The studies for the initially slowly +/- chirped pulses (with the chirp coefficients C= -0.83 to +0.83) show that only the pulse peak intensity and spectral bandwidth are important for the process, and initial phase of pulses does not impact to the process. Our studies curried out for the Gaussian, secant-hyperbolic and super-Gaussian pulses, in the range of parameters R=0.25 to 1 and $\zeta \approx 1$ to 20000, show up to 30^X SC.

References

[1] Mouradian L, Zohrabyan A, Villeneuve A, Yavrian A, Rousseau G, Piche M, Froehly C, Louradour F and Barthélémy A 2000 *CLEO / Europe'2000, Conference Digest, CTuH6* **39**
[2] Akhmanov S A, Vysloukh V A and Chirkin A S 1992 *Optics of Femtosecond Laser Pulses,* (New York: AIP)
[3] Agrawal G P 2001 *Nonlinear Fiber Optics* (Academic, third ed.)
[4] Kutuzyan A, Mansuryan T, Kirakosian A and Mouradian L 2003 *Proc. SPIE* **5135** 156
[5] Hasegawa A and Tappert F 1973 *Appl. Phys. Lett.* **23** 142
[6] Mollenauer L, Stolen R and Gordon J 1980 *Phys. Rev. Lett.* **45** 1045
[7] Mollenauer L, Stolen R and Gordon 1986 *J Opt. Lett.* **11** 289

[8] Tognetti M and Crespo H 2007 *J. Opt. Soc. Am.* B **24** 1410
[9] Amorim A, Tognetti M, Oliveira P, Silva J, Bernardo L, Kärtner F and Crespo H 2009 *Opt. Lett.* **34** 3851
[10] Salem A, Cherif R and Zghal M 2011 *Opt. Expr.* **19** 19995
[11] Yesayan G 2012 *J. Contemp. Phys.* **47/5** 225
[12] Toneyan H, Zeytunyan A, Zadoyan R and Mouradian L *J. Phys.: Conf. Ser.* submitted (2015)
[13] Hardin R and Tappert F *1973 SIAM Rev. Cronicle* **15** 423
[14] Fisher R and Bischel W 1973 *Appl. Phys.Lett.* **23** 661

Classic, all-fiber, and similaritonic techniques of spectral compression

H Toneyan[1,2], A Zeytunyan[1,3], R Zadoyan[3] and L Mouradian[1]

[1] Ultrafast Optics Laboratory, Faculty of Physics, Yerevan State University, Armenia
[2] CANDLE Synchrotron Research Institute, Yerevan, Armenia
[3] Newport Technology & Applications Centre, Irvine, CA 92606, USA

E-mail: h.toneyan@gmail.com

Abstract. A comparative experimental study of five techniques of spectral compression based on the self-phase modulation and phase addition processes, similariton generation and soliton effect is reported. In view of all-fiber configuration, the hollow-core fiber was used in experiments along with the standard single-mode one. The nonlinear process of spectral self-compression, a spectral analogue of soliton effect compression, in a hollow core fiber with anomalous dispersion is observed experimentally. To reveal the nature and origin of aberrations of spectro-temporal lenses induced by the phase modulation and phase addition process, the experiments are conjugated with numerical studies. An up to 23x aberration free spectral focusing in the time lens induced by similariton through sum-frequency generation is demonstrated experimentally.

1. Introduction

The interest to the nonlinear process of spectral compression in a dispersive delay line followed by a single-mode fiber, revealed over two decades ago [1-4], is motivated by the numerous prospective applications in ultrafast optics and photonics, often based on the spatiotemporal analogy and the concept of time lens [5,6]. Spectral compression is a temporal analogue of diffracted beam collimation in a light-induced lens. In the dispersive delay line, pulses are stretching and phase-modulating (frequency-chirping). Further compensation of the dispersively accumulated phase-shift by means of self-phase modulation in a nonlinear fiber leads to the spectral narrowing: the temporal phase-shift induced by the Kerr effect in the fiber, like a time lens, "collimates" the radiation in time, and "focuses" the spectrum. The time lens, like the spatial one, has a more general feature of Fourier transformation, leading to the conversion of temporal information to the spectral domain, and thus to the applications to the signal analysis and synthesis problems in ultrafast optics: spectral imaging of the pulse temporal profile for direct femtosecond pulse measurements [7,8] and fine frequency tuning of radiation along with spectral compression [8] for resonant spectroscopy, generation of dark solitons [5], and the femtosecond pulse undistorted delivery [9]. Spectral compression was recently proposed also for similariton fiber laser architecture instead of highly dissipative spectral filtering [10]. Spectral compression is of interest in the context of nonlinear vibrational microscopy where high spectral brightness is required, such as in coherent anti-Stokes Raman scattering (CARS) or stimulated Raman scattering microscopy. 8.7x spectral compression through the generation of negatively chirped parabolic pulses was demonstrated in [11], showing aberration-free spectral focusing due to the optimal parabolic pulse shape. Further development resulted in 12x compression in an all-fiber configuration for telecommunication wavelengths, utilizing kilometre-long fiber [12].

Current problems of ultrafast optics and photonics, particularly femtosecond signal generation, manipulation, delivery and characterization, stimulate a detailed study of the spectral compression process and the variety of the its modern techniques. We comparatively studied three spectral compression schemes in our initial experiments [13]. In this work, we complete this study and present five spectral compression techniques, together with numerical simulations. We pay special attention to aberration of spectral focusing in view of achievement aberration-free process, as well as to all-fiber configuration of the process implementation.

2. Experiment

We report on a comparative study of spectral compression (SC) techniques based on the self-phase modulation (SPM) and sum-frequency generation (SFG) processes, similariton generation and soliton effect. In our experiments, we use the radiation of a standard Coherent Verdi 10 + Mira 900F femtosecond laser system with the following parameters: 100 fs FWHM pulse duration, ~ 11 nm spectral bandwidth, 800 nm central wavelength, 1.5 W average power at the 76 MHz repetition rate.

2.1. Classic technique of spectral compression

First, we experiment the SC in the classic scheme, i.e. in a system consisting of prism compressor as a dispersive delay line (DDL) with negative dispersion, and a single-mode fiber (SMF) as a nonlinear medium for SPM. Our DDL consisted of a 3.75 m separated SF11 prism pair with the reverse mirror. A 92 cm-long Newport F-SE 780 nm SMF was used for nonlinear SFM (figure 1a).

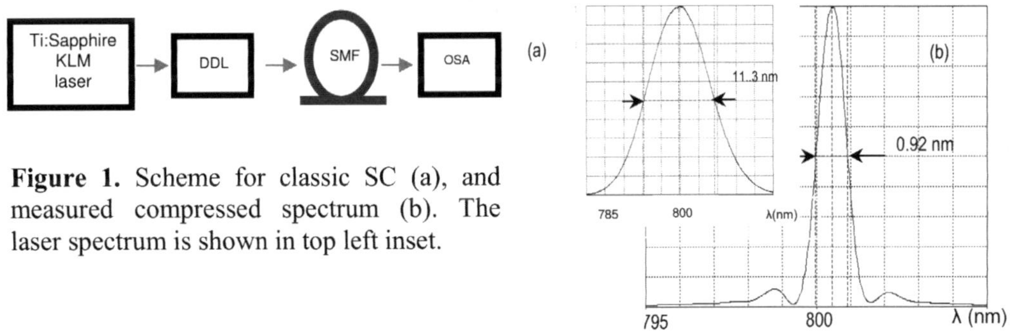

Figure 1. Scheme for classic SC (a), and measured compressed spectrum (b). The laser spectrum is shown in top left inset.

Starting from the 11.3 nm-bandwidth pulses, 12.3x spectrally compressed pulses with 0.92 nm bandwidth are obtained at the output (figure 1b), with ~ 2 ps autocorrelation duration and 0.42 W average power. Such spectrally compressed pulses can be useful for the CARS techniques.

2.2. Spectral self-compression

Thereafter, we have experimented hollow core fiber (HCF, ThorLabs HCF-800, 2 m) with anomalous dispersion at 800 nm wavelength. First, we simply coupled the radiation into HCF and measured the output spectrum depending on radiation power in the fiber (figure 2a).

Figures 2b and 2c demonstrate up to 1.3x spectral narrowing in HCF as a result of combined impact of weak nonlinearity and anomalous dispersion, evidencing the process of self-SC, which is the spectral analogue of soliton compression [14]. Detailed numerical studies of the process based on the solution of nonlinear Schrödinger equation, with the factors of Kerr-nonlinearity and anomalous dispersion, show few-tens of ratios of spectral self-compression [15]. Figure 2c shows nonlinear character of the process: both the spectral bandwidth and pulse autocorrelation duration at the output of HCF are decreasing monotonically with the increase of coupled radiation power. The spectral shift of SC radiation visible in figure 2b is caused by high-order dispersive effects, in accordance with our

Journal of Physics: Conference Series **672** (2016) 012016 doi:10.1088/1742-6596/672/1/012016

numerical analysis. Thus, due to the combined impact of low nonlinearity and anomalous dispersion, we have experimentally observed spectral self-compression in a hollow-core fiber.

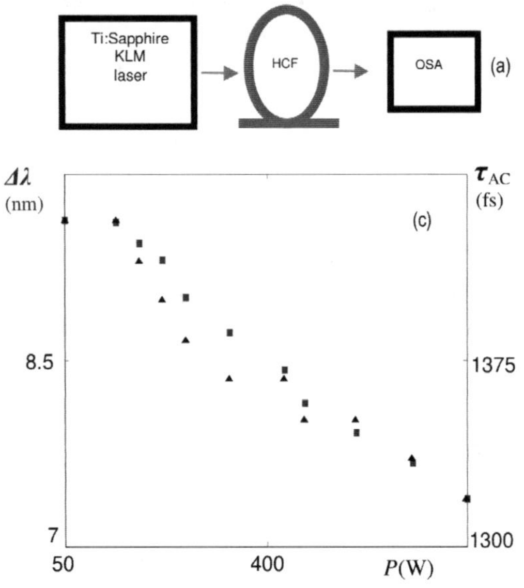

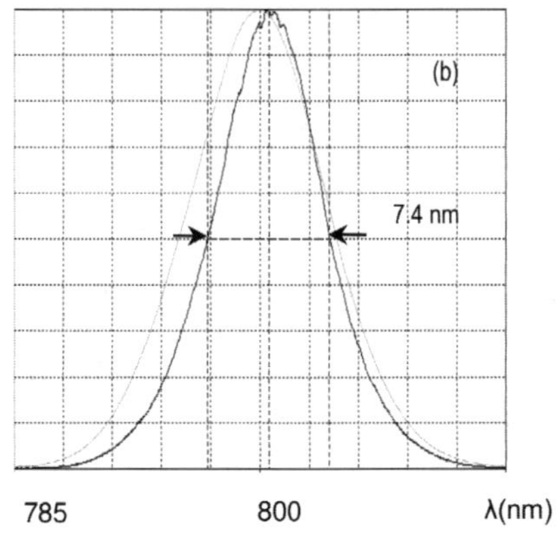

Figure 2. Scheme of self-SC in HCF (a), and measured spectra at the input (pink) and output (black) of HCF (b); the measured spectral bandwidths (blue squares) and pulse autocorrelation duration (black triangles) versus coupling power (c).

2.3. All-fiber spectral compression

Afterwards, the output pulse from the HCF was coupled into a conventional SMF (Newport F-SE @ 780 nm, 80 cm) (figure 3a). We have obtained compression of the initial 10.9 nm spectrum down to 1.3 nm (figure 3b), corresponding to 8.4x SC, with 0.4 W average output power. The strong side lobes of the compressed spectrum observed in figure3b, are caused, according to our numerical simulations, by the high-order dispersion of HCF, measured via spectral interferometry. In this experiment, the HCF dispersion was tuned and SC was optimized by setting the laser wavelength to 808 nm. Further development in this direction is anticipated by using the HCFs with flattened dispersion, as well as by splicing the HCF and SMF.

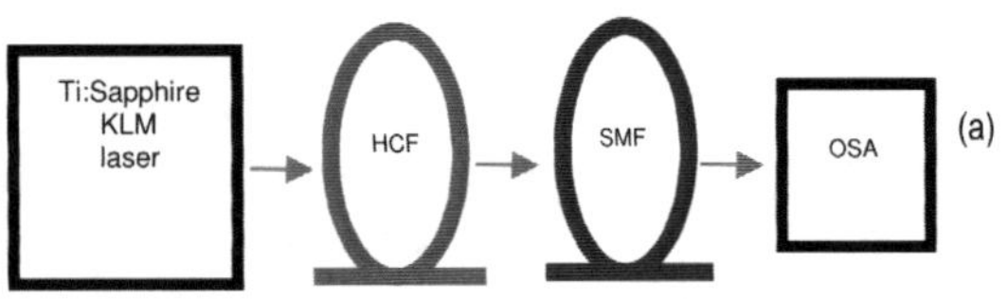

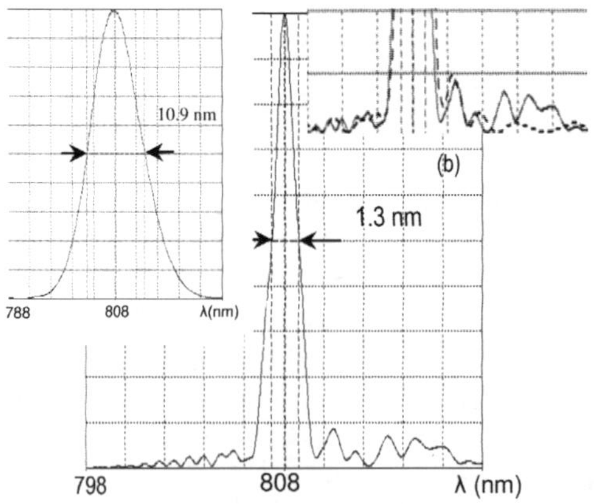

Figure 3. All-fiber scheme of SC with HCF as DDL (a) and compressed spectrum with laser spectrum at top left and numerical simulation of aberrations at the right corner (b).

2.4. Similaritonic spectral compression

Thereafter, we implemented SC through a phase addition process, particularly through the SFG [8]. Figure 4 illustrates these studies. In this experiment, we have split the initial laser radiation into high- and low-power beams (80% and 20%). We coupled the high-power beam to standard SMF (1.65 m Newport F-SPF PP@ 820 nm), and generated a 100 nm bandwidth nonlinear-dispersive similariton [16]. The low-power beam was directed to DDL consisting of a 3.5 m separated SF11 prism pair with a reverse mirror. Then, the beams were focused in a BBO crystal for SFG, resulting in 23.3x SC for frequency, and more than 90x SC for wavelength, down to 0.12 nm bandwidth (figure 4b). The SFG-SC ratio $\Delta\omega_{in}/\Delta\omega_{out}$ corresponds to the $\Delta\lambda_{in}(400nm)=\Delta\lambda_{in}(800nm)/4$ bandwidth of the SFG spectrum measured in the absence of DDL. An important benefit of similaritonic technique is the opportunity of frequency tuning of signal radiation in a range given by the similariton spectrum. The latter is caused by the chirp of similariton, which allows shifting the SFG spectrum controlling the delay between similariton reference and signal pulses.

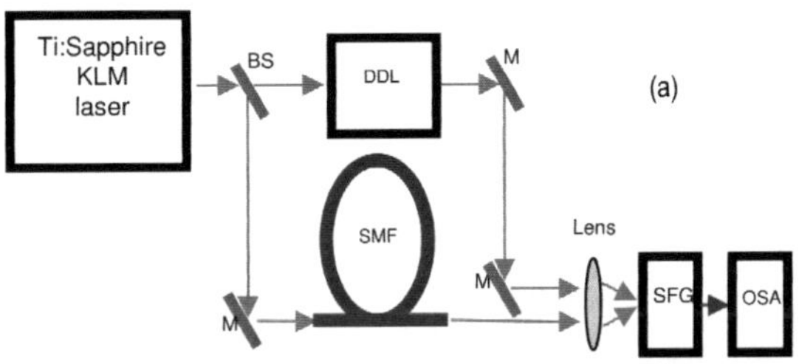

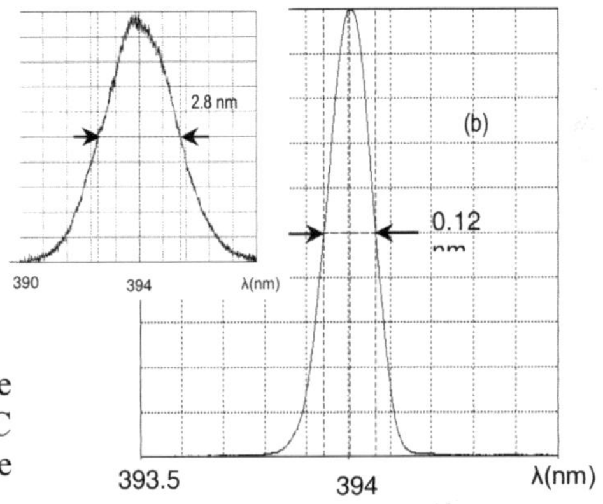

Figure 4. Experimental setup of SFG-SC (a). The spectrum at the system output shows the 23.3x SC (b). The inset of (b) is the SFG-spectrum in absence of the DDL (reference measurement).

2.5. SFG-spectral compression with HCF

We have also carried out experiments with HCF in the SFG scheme, replacing the prism DDL of similaritonic technique by a 2 m-long HCF (figure 5a). As a result we had compression from 10 nm at 800 nm central wavelength down to 0.2 nm on second harmonic wavelength, which corresponds to 11x SC for frequencies, and 44x SC for wavelengths (figure 5b). Although the original similaritonic technique of Section 2.4 is aberration free, there are side lobes for compressed spectrum (figure 5b) in this configuration, caused by uncompensated higher order dispersion of HCF.

3. Simulations

To understand the nature and origins of the side lobes in SC configurations with HCF, we have done numerical studies towards SC with uncompensated higher orders of dispersion (figure 3b). For the pulse propagation in the SMF, a standard nonlinear Schrödinger equation with the terms of Kerr nonlinearity and GVD is solved by the use of split-step Fourier method [16-18].

First, we tested DDL with only second order of dispersion, which resulted in spectral compression represented in fig.6a (red curve stands for initial spectrum, blue curve for compressed spectrum). Thereafter we changed the DDL in the program to have second, third and fifth orders of dispersion and modelled spectral compression (figure 6c). We tested third and fifth orders of dispersion which resulted in SC represented in green and brown curves (figure 6b) respectively. Thereafter we have tested the combined impact of both effects (figure 6c). We found out that side lobes similar to lobes in spectral compression scheme with use of HCF (figure 3b) occur when we put considerable values of third and fifth orders of dispersion, which are not compressed in self-phase modulation process. We assume, based on this numerical studies, that side lobes are caused by third and fifth orders of dispersion. Forth order of dispersion was also tested but had no considerable impact on spectrum deformations.

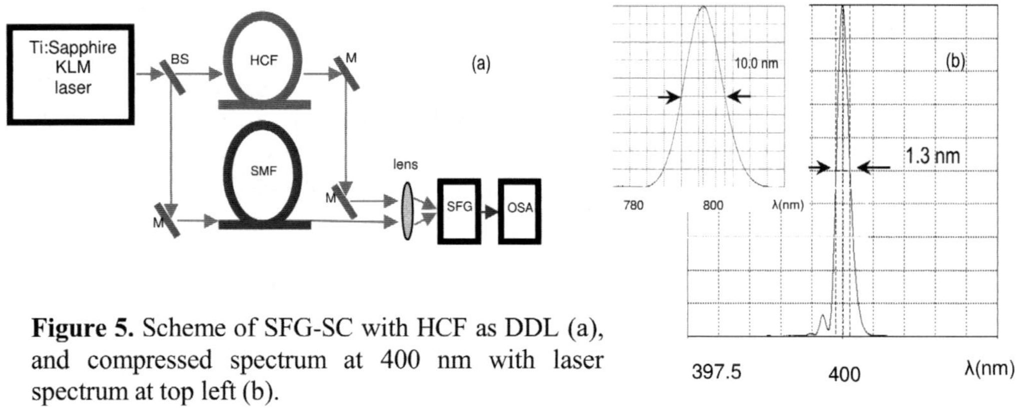

Figure 5. Scheme of SFG-SC with HCF as DDL (a), and compressed spectrum at 400 nm with laser spectrum at top left (b).

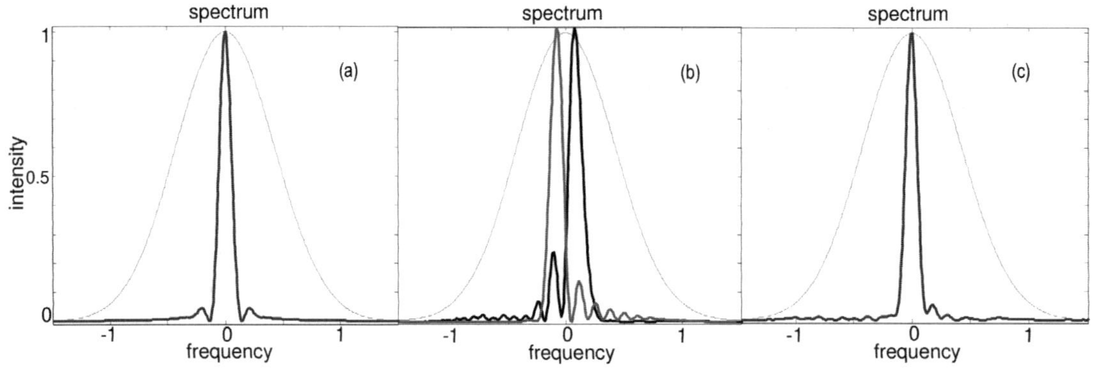

Figure 6. Results of numerical simulations for SC. Red curves are for inita spectrum. (a) Blue curve is compressed spectrum considering 2nd order dispersion only corresponding to classic scheme of Sec. 2.1. (b) SC with 2nd + 3rd order dispersion (green), and 2nd +5th order dispersion (black). (c) SC with the combined impact of 2nd +3rd +5th order dispersion corresponding to all fiber scheme with HCF of Sec. 2.3.

Concluding, we have implemented femtosecond pulse SC through five different techniques: classic and all-fiber through SPM, similaritonic and all-fiber similaritonic through SFG. Spectral compression with the ratios 12.3x, 8.4x, and 11x for the classic, all-fiber, and all-fiber similaritonic techniques, correspondingly, are achieved. Spectral focusing in all this cases require compensation of high-order dispersion to avoid side lobes in compressed spectra.

We demonstrated a 23.3x aberration-free SC for the similaritonic SFG-method.

Finally, we have experimentally observed 1.3x spectral narrowing in HCF as a result of combined impact of weak nonlinearity and anomalous dispersion, proving the process of self-SC, which is the spectral analogue of soliton effect compression.

References

[1] Muradyan L Kh, Markaryan N L, Papazyan T A and Ohanyan A A 1990 *in Conference on Lasers and Electro-Optics (OSA Tech. Digest Series)* **7** CTUH32
[2] Margaryan N L, Mouradian L Kh and Papazyan T A 1991 *Sov. J. Quant. Electron.* **21** 783–785
[3] Oberthaler M and Hopfel R A 1993 *Appl. Phys. Lett.* **63** 1017–1019
[4] Washburn B R, Buck J A and Ralph S E 2000 *Opt. Lett.* **25** 445–447
[5] Mouradian L Kh, Zohrabyan A V, Villeneuve A, Yavrian A, Rousseau G, Piche M, Froehly C, Louradour F and Barthélémy A 2000 *CLEO-Europe, Conf. Digest (OSA Trends in Optics and Photonics)* **39** CTuH6
[6] Salem R, Foster M A and Gaeta A L 2013 *Adv. Opt. Photon.* **5** 274–317
[7] Mouradian L Kh, Louradour F, Messager V, Barthélémy A and Froehly C 2000 *IEEE J. Quantum Electron.* **36** 795–801
[8] Mansuryan T, Zeytunyan A, Kalashyan M, Yesayan G, Mouradian L, Louradour F and Barthélémy A 2008 *J. Opt. Soc. Am.* B **25** A101–A110
[9] Clark S W, Ilday F Ö and Wise F W 2001 *Opt. Lett.* **26** 1320–1322
[10] Boscolo S, Turitsyn S K and Finot C 2012 *Opt. Lett.* **37** 4531-4533
[11] Andresen E R, Dudley J M, Oron D, Finot C and Rigneault H 2011 *Opt. Lett.* **36** 707–709
[12] Fatome J, Kibler B, Andresen E R, Rigneault H and Finot C 2012 *Appl. Opt.* **51** 4547-4553
[13] Toneyan H, Zeytunyan A, Mouradian L, Tsakanov V, Louradour F, Barthelemy A and Zadoyan R 2014 *Frontiers in Optics 2014* FW4D.5
[14] Grigoryan A P, Yesayan G L, Kutuzyan A A and Mouradian L Kh 2015 *Adv. Sci. Focus* (submitted)
[15] Grigoryan A, Yesayan G, Kutuzyan A and Mouradian L 2015 *J. Phys.: Conf. Ser.* (submitted)
[16] Zeytunyan A, Muradyan A, Yesayan G, Mouradian L, Louradour F and Barthélémy A 2011 *Opt. Commun.* **284** 3742–3747
[17] Akhmanov S A, Vysloukh V A and Chirkin A S 1992 *Optics of Femtosecond Laser Pulses* (AIP)
[18] Agrawal G P 1995 *Nonlinear Fiber Optics* (Academic)

Carrier localization, Anderson transitions and stripe formation in hole-doped cuprates

S Dzhumanov and Z S Khudayberdiev
Institute of Nuclear Physics AS RUz, 100214, Tashkent, Uzbekistan
E-mail: `zafar@inp.uz`

Abstract. Three distinctly different scenarios are proposed for the carrier localization in three-dimentional (3D) lightly doped cuprates in which the self-trapping and pairing of hole carriers (i) near the small-radius dopants and (ii) in a defect-free deformable lattice lead to the formation of the extrinsic and intrinsic (bi)polaronic states in the charge-transfer gap of the cuprates, and (iii) the self-trapping of hole carriers away from the large-radius dopants results in the formation of the in-gap hydrogenic impurity states. We have shown that the extrinsic and intrinsic 3D large bipolarons exist in La-based lightly doped cuprates at $\eta = \varepsilon_\infty/\varepsilon_0 < 0.127$ and $\eta < 0.138$, respectively, where $\varepsilon_\infty(\varepsilon_0)$ is the optic (static) dielectric constant. We use the uncertainty relation to obtain the specific conditions for the Anderson and new MITs in cuprates. The applicability limits of these MITs in La-based cuprates are clarified. Our results are in good agreement with the existing experiments on La-based cuprates.

1. Introduction

Understanding the mechanisms of carrier localization, metal-insulator transitions (MITs), stripe formation and suppression of high-T_c superconductivity in hole-doped cuprates remains one of the central issues in condensed matter physics. In hole-doped cuprates, the phenomena of carrier localization and MITs seem to be much more complicated by carrier-dopant (defect)-lattice or carrier-lattice interaction effects, which usually were ignored in the existing theoretical models. In reality, there are important unexplained differences between the MITs observed in lightly doped [1, 2, 3], underdoped [4, 5, 6] and optimally doped [7] cuprates. Another unresolved issue in the physics of high-T_c cuprates is the role of the electronic inhomogeneity and charge ordering in the phase separation in the form of alternating dynamic and static stripes [8, 9, 10, 11, 12], which is intimately related to carrier localization, MITs and superconductivity in these materials. Our main purpose is to understand the possible microscopic mechanisms leading to the carrier localization, MITs, stripe formation and suppression of superconductivity and to propose a unified theoretical description of these interrelated phenomena in the cuprates.

2. Possible Scenarios for Carrier Localization in Lightly Doped Cuprates

Hole doping of the cuprates produces first quasi-free holes having the mass m_b in the oxygen valence band. In polar cuprates, these hole carriers interacting both lattice vibrations and with lattice defects (e.g. dopants or impurities) are self-trapped either near the defects (defect-assisted extrinsic self-trapping) or in a defect-free deformable lattice (phonon-assisted intrinsic self-trapping). Therefore, the ground states of such hole carriers are their self-trapped (i.e.

localized extrinsic and intrinsic polaronic) states lying in the CT gap of the cuprates [13]. A large ionicity of the cuprates $\eta = \varepsilon_\infty/\varepsilon_0 << 1$ (where ε_∞ and ε_0 are the high-frequency and static dielectric constants, respectively) enhances the polar electron-phonon interaction and the tendency to polaron formation. Actually, the relevant charge carriers in hole-doped cuprates are large polarons [13, 14, 15, 16] and the strong electron-phonon interactions are responsible for enhancement of the polaron mass $m_p = (2.0 - 3.0)m_b$ [17] (where $m_b \simeq m_e$ is the free electron mass). The formation of nearly small Frölich polaron [18] might be also relevant. According to Ref.[13], the ground state energies or binding energies E_{ep} of extrinsic polarons would increase rapidly with decreasing ε_∞ from 5 to 3 or with increasing η from 0 to 0.12 and are equal to $E_{ep} \simeq (0.11 - 0.18)$ eV (for $\varepsilon_\infty = 3.5 - 4.5$ and $\eta = 0.12$) and $E_{ep} \simeq (0.086 - 0.14)$ eV (for $\varepsilon_\infty = 4$ and $\eta = 0 - 0.12$). Whereas the binding energies E_p of intrinsic polarons would decrease noticeably with increasing η from 0 to 0.12 (i.e. $E_p \simeq (0.085 - 0.065)$ eV for $\varepsilon_\infty = 4$ and $\eta = 0 - 0.12$), but E_p increases from 0.054 eV to 0.09 eV with decreasing ε_∞ from 4.5 to 3.5 at $\eta = 0.10$. We believe that the carrier-defect-phonon and carrier-phonon interactions together with the charge inhomogeneities play an important role in hole-doped cuprates and are responsible for the carrier localization and segregation. In these materials, the inhomogeneous spatial distribution of polaronic carriers leads to their segregation into carrier-rich and carrier-poor regions. Perhaps the carrier-defect-phonon and carrier-phonon interactions give rise to charge aggregation in carrier-rich metallic regions together with charge depletion in spatially separated carrier-poor regions with no mobile carriers. In general, the local charge inhomogeneity and the competition between the kinetic energy and the aligning interactions produce nanoscale self-organized structures called stripes. Further, the anisotropy of the dielectric constants (ε_∞ and ε_0) and smallness of η in the cuprates favor such a carrier segregation.

3. Possible mechanisms of metal - insulator transitions and stripe formation in cuprates

The binding energy of the intrinsic large bipolaron is defined as

$$\Delta_b = E_{bB} - 2\varepsilon_F, \tag{1}$$

At a certain doping level $n = n_c$ or $x = x_c = n_c/n_a$, $\Delta_b = 0$ and the large bipolaron will dissociate into two large polarons. Therefore, the critical carrier concentration n_c determined as

$$n_c = (m_p E_{bB})^{3/2}/3\pi^2\hbar^3 \tag{2}$$

For the LSCO system we can evaluate n_c using the parameter values $m_p = 2.1m_e$, $\varepsilon_\infty = 3.5-4.5$, $\eta = 0.04 - 0.08$, $E_{bB} \approx 0.01 - 0.10$ eV. We attempt to find the conditions under which such a insulator-to-metal transition occurs in doped materials.

The conditions for carrier localization or delocalization can be obtained by using the uncertainty principle: $\Delta p \Delta x \geq \hbar/2$. This uncertainty relation can be written as

$$\Delta x \cdot \Delta E \simeq \frac{\hbar^2(\Delta k)^2}{2m^*}\frac{1}{2\Delta k}, \tag{3}$$

Taking into account that in the impurity band the uncertainties in the energy of carriers is of order $W_I/2$ and the uncertainty in their wave vector is about $1/a_I$, the relation (3) can be rewritten in the form (cf.[19])

$$\Delta x \cdot \Delta E \cong W_I a_I/4 \tag{4}$$

Since, a random potential V_0 leads to the uncertainties in the energy $\Delta E \sim V_0/2$ and coordinate $\Delta x \sim a_H$ of carriers, the condition for Anderson localization can be written as

$$V_0/W_I > 0.5a_I/a_H, \tag{5}$$

Journal of Physics: Conference Series **672** (2016) 012017 doi:10.1088/1742-6596/672/1/012017

We believe that the strong carrier-defect-phonon and carrier-phonon interactions can initiate MITs in cuprates. The condition for carrier localization or a new type of MIT can be written as

$$E_{pI}/W_I > 0.25a_I/R_I, \tag{6}$$

from which it follows that the MIT is governed by the ratios E_{pI}/W_I and a_I/R_I.

We examine below the possibility of the Anderson and new MITs and stripe formation in La-based cuprates.

3.1. The Anderson-type metal-insulator transitions

In inhomogeneous hole-doped cuprates the randomness in dopant distribution affects on the depths of the potential wells of impurities. In disorder system the volume Ω_0 may contain two or more impurities at the random spatial distribution of dopants and the energy of a carrier in the field of l impurities is equal to lE_I [20]. One can expect that the depth of the potential well for the carrier varies from 0 to lE_I and $V_0/2 \simeq lE_I/2$. The case $l = 2$ is assumed to be more probable. Therefore, we can take $V_0 = 2E_I^H$ and write the condition (5) in the form

$$z \exp\left[-R/a_H\right] = a_H/a_I \tag{7}$$

Taking into account that dopants form different superlattices with the site disorder, we obtain the following criteria for the Anderson-type MITs:

$$(z/n^{1/3}a_H)\exp[-1/n^{1/3}a_H] = 1 \quad \text{for} \quad z = 6, \tag{8}$$

$$(\sqrt[3]{2}\cdot z/n^{1/3}a_H)\exp[-\sqrt{3}/\sqrt[3]{4}\cdot n^{1/3}a_H] = 1 \quad \text{for} \quad z = 8, \tag{9}$$

$$(\sqrt[3]{4}\cdot z/n^{1/3}a_H)\exp[-2^{1/6}/n^{1/3}a_H] = 1 \quad \text{for} \quad z = 12. \tag{10}$$

From equations (8),(9) and (10), we find $n^{1/3}a_H \simeq 0.3530$, $n^{1/3}a_H \simeq 0.3148$ and $n^{1/3}a_H \simeq 0.2618$, respectively. If we assume $a_H = 7 - 8Å$ [1], we find $x_c \simeq 0.0116 - 0.0173$ (for $z = 8$), and $x_c \simeq 0.0163 - 0.0243$ (for $z = 6$) at which the coexistence of the metallic and insulating stripes and the transition from the metallic to the insulating behavior were observed in lightly doped cuprates (see [21, 22]). For example, experimental results show that the degeneration of hole carriers occurs at the appearance of the first metallic stripes for $x \approx 0.02$ [2].

3.2. The new metal-insulator transitions and stripe formation

The width of the energy band of extrinsic large polarons or impurities with tightly bound large polarons can be defined as (see [23])

$$W_I = 2z\alpha^3 v_0 D_0 \exp\left[-\alpha R\right] \tag{11}$$

If v_0 would be of the order of R_I^3, the quantity $\alpha^3 v_0$ may be replaced by unity. The D_0 is of the order of E_{pI}. Thus, we obtain the following criteria for the new MITs from the relation (6):

$$(0.5z/n^{1/3}R_I)\exp[-1/n^{1/3}R_I] = 1 \quad \text{for} \quad z = 6,$$

$$(0.5\sqrt[3]{2}\cdot z/n^{1/3}R_I)\exp[-\sqrt{3}/\sqrt[3]{4}\cdot n^{1/3}R_I] = 1 \quad \text{for} \quad z = 8,$$

$$(0.5\sqrt[3]{4}\cdot z/n^{1/3}R_I)\exp[-2^{1/6}/n^{1/3}R_I] = 1 \quad \text{for} \quad z = 12,$$

from which it follows that $n^{1/3}R_I \simeq 0.651$ (for $z = 6$), $n^{1/3}R_I \simeq 0.453$ (for $z = 8$), $n^{1/3}R_I \simeq 0.3354$ (for $z = 12$). We are now able to evaluate n_c in La-based cuprates using the above values of $n_c^{1/3}R_I$. For $\varepsilon_\infty = 4.5$, and $\eta = 0.04 - 0.12$ the value of R_I is found to vary between 8.44 and 9.91 Å. In these cases, the MITs and stripe formation occur at $x_c \simeq 0.0539 - 0.0872$ (for $z = 6$) and $x_c = 0.0182 - 0.0294$ (for $z = 8$) in La- based cuprates with small-radius dopants. When $\varepsilon_\infty = 3.5$, $\eta = 0.04 - 0.12$ and $R_I \simeq 6.41 - 7.56$, the MITs and stripe formation occur in these systems at $x_c \simeq 0.1213 - 0.1990$ (for $z = 6$) and $x_c \simeq 0.0409 - 0.0671$ (for $z = 8$). For $\varepsilon_\infty < 3.5$ and $z = 6$, the metal-insulator boundary of La-based cuprates containing small-radius dopants lies in the deeply overdoped region ($x_c > 0.2$)

4. Conclusions

In this work we have studied the possible mechanisms of carrier localization, charge ordering, MITs and stripe formation and propose a unified theoretical description of these interrelated phenomena in inhomogeneous hole-doped cuprates. The phenomena of self-trapping and pairing of hole carriers in the vicinity of the dopants (impurities) and in a defect-free deformable lattice at their interaction with impurities, acoustic and optical lattice vibrations in 3D cuprate compounds have been investigated quantitatively using the continuum model of ionic crystal and adiabatic approximation. These distinctly different scenarios for carrier localization lead to the formation of the extrinsic large (bi)polaronic states, the hydrogenic impurity states (i.e. impurities with loosely bound free carriers or large polarons) and the intrinsic large (bi)polaronic states in the CT gap of the cuprates. The binding energies and sizes of the extrinsic and intrinsic large polarons and bipolarons are calculated variationally, taking into account the short- and long-range parts of the carrier-defect-phonon and carrier-phonon interactions. Using the uncertainty relation, we are able to obtain the specific conditions for the MITs caused by disorder, carrier-dopant-phonon and carrier-phonon interactions. Validity of the Anderson and new MITs for the La-based cuprates is clarified. Our results have shown that the Anderson MITs may occur in lightly doped cuprates containing large-radius dopants, while the new MITs take place in deeply underdoped ($x \approx 0.02 - 0.05$), slightly underdoped ($x \lesssim 0.13$), optimally doped ($x \gtrsim 0.15$) and deeply overdoped ($x > 0.2$) cuprates containing both the small-radius dopants and the large-radius dopants. Our results consistent with the experimental data on carrier localization, MITs and stripe formation in La-based cuprates.

References

[1] Kastner M A, Birgeneau R J, Shirane G and Endoh Y 1998 *Rev. Mod. Phys.* **70** 897
[2] Ichikawa N, Uchida S, Tranquada J M, Niemöller T, Gehring P M, Lee S-H and Schneider J R 2000 *Phys. Rev. Lett.* **85** 1736
[3] Ishida K, Aya H, Tokunaga Y, Kotegawa H, Kitaoka Y, Fujita M and Yamada K 2004 *Phys. Rev. Lett.* **92** 257001
[4] Fink J *et al* 1991 *Physica C* **45** 185-189
[5] Ono S, Ando Y, Murayama T, Balakirev F F, Betts J B and Boebinger G S 2001 *Physica C* **357-360** 138
[6] Ando Y, Ono S, Sun X F, Takeya J, Balakirev F F, Betts J B, and Boebinger G S 2004 *Phys. Rev. Lett.* **92** 247004
[7] Boebinger G S *et al* 1996 *Phys. Rev. Lett.* **77** 5417
[8] Tranquada J M, Sternlieb B J, Aye J D, Nakamura Y and Uchida S 1995 *Nature* **375** 561
[9] Zaanen J, Osman O Y and van Saarloos W 1998-II *Phys. Rev.* B **58** R11868
[10] Castellani C, Castro C Di and Grilli M 1998 *J. Phys. Chem. Solids.* **59** 1694
[11] Kivelson S A, Bindloos I, Fradkin E, Oganesyan V, Tranquada J, Kapitulnik A and Howard C 2003 *Rev. Mod. Phys.* **75** 1201
[12] Kato T, Noguchi T, Saito R, Machida T and Sakata H 2007 *Physica C* **460-462** 880
[13] Dzhumanov S, Baimatov P J, Ganiev O K, Khudayberdiev Z S, Turimov B V 2012 *J. Phys. Chem. Solids* **73** 484
[14] Kastner M A, Birgeneau R J, Shirane G and Endoh Y 1998 *Rev. Mod. Phys.* **70** 897
[15] Bi X X and Eklund P C 1993 *Phys. Rev. Lett.* **70** 2625
[16] Emin D 2013 *Polarons* (Cambridge: Cambridge University Press)
[17] Basov D N and Timusk T 2005 *Rev. Mod. Phys.* **77** 721
[18] Alexandrov A S and Kornilovitch P E 1999 *Phys. Rev. Lett.* 82 807; Alexandrov A S and Yavidov B Y 2004 *Phys. Rev. B* **69** 073101

Journal of Physics: Conference Series **672** (2016) 012017 doi:10.1088/1742-6596/672/1/012017

[19] Ridley B K 1986 *Quantum Processes in Semiconductors* (Moscow: Mir)
[20] Shklovskii B I and Efros A L 1979 *Electronic Properties of Doped Semiconductors* (Moscow: Nauka)
[21] Komiya S, Ando Y, Sun X F and Lavrov A N 2002 *Phys. Rev.* B **65** 214535
[22] Anshukova N V, Golovashkin A I, Ivanova L I and Rusakov A P 2003 *Zh. Eksp. Teor. Fiz.* **123** 1188
[23] Mott N F and Devis E A 1974 *Electronic Proceses in Non-Crystaline Materials* (Moskow: Mir)

Sapphire shaped crystals for medicine

I A Shikunova and V N Kurlov

Institute of Solid State Physics RAS, Chernogolovka, 142432, Russia

E-mail: kurlov@issp.ac.ru

Abstract. The favorable combination of excellent optical and mechanical properties of sapphire makes it an attractive structural material for medicine. We have developed a new kind of medical instruments and devices for laser photodynamic and thermal therapy, laser surgery, fluorescent diagnostics, and cryosurgery based on sapphire crystals of various shapes with capillary channels in their volume.

1. Introduction

Sapphire has a high refractive index and a broad transmission band spanning the UV, visible, and IR bands. Sapphire also has high hardness, very good thermal conductivity, tensile strength, and thermal shock resistance. This favorable combination together with chemical inertness and biocompatibility, makes it an attractive structural material for medicine. A number of growth techniques have been developed to manufacture sapphire crystals of complex shapes directly from melt [1]. Sapphire shaped crystals holding multiple capillary channels with diameters are less than 600 microns in the crystal volume have been grown. This development has given rise to the emergence of essentially new areas of application.

In recent years we are developing new instruments and devices for laser treatment on the base of new sapphire shaped crystals:

- Sapphire needle capillary irradiators for laser radiation delivery for photodynamic interstitial therapy, laser thermotherapy and an ablation of tumors of a prostate, a liver, a thyroid gland, etc.
- Sapphire scalpels with possibility of coagulation and fluorescent diagnostics of malignancy of resected tissues in a vicinity of a cutting edge directly in the surgery process;
- Cryoapparatus for destruction with sapphire cryotips-ligthguide;
- System for removal of brain tumors based on a sapphire probe with demarcation of borders of a tumor by fluorescent diagnosis and others.

Sapphire needle capillaries were developed as new laser waveguide introducers for delivery radiation into a tumour during interstitial laser photodynamic therapy, thermotherapy, and ablation of tumors [2]. These needles allow one to increase the irradiation volume substantially, to obtain an optimal temperature distribution, to simplify the design, and to eliminate a system for cooling the device. The high hardness of sapphire provides a stable point on the irradiator end for independent introduction of the irradiator into the tissue without using directors that lead to the increase in irradiator cross-section. The use of sapphire irradiators make it possible to improve the control over the dynamics of spatial photothermal distribution during the whole irradiation procedure, since the effective redistribution of released heat by the sapphire decreases the possibility of formation of overheating nuclei, leading to the appearance of thrombi, nontransparent to laser radiation.

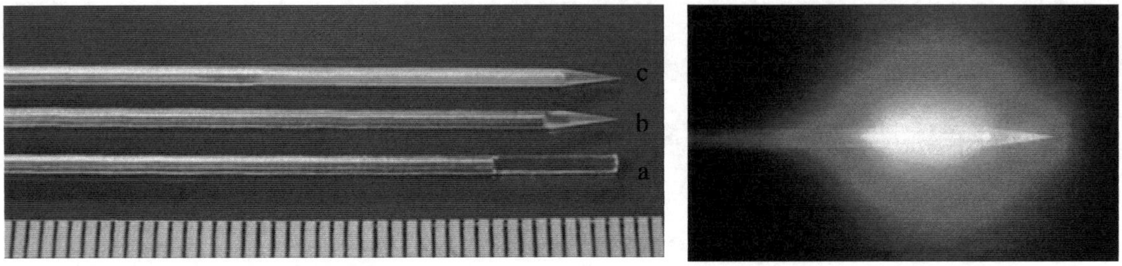

Figure 1. At the left: sapphire needle capillaries (external diameter of 1.2 mm, internal diameter of 0.5 mm): (a) as grown closed capillary; (b) capillary with a point formed by mechanical operation of the butt; (c) capillary with a point and diffuser formed by mechanical operation. At the right: sapphire needle combined with quartz fiber (the geometry of light field).The geometry of light field in sapphire needle (OD 1.2 mm, ID 0.5 mm) combined with quartz fiber.

Sapphire scalpels with simultaneous incision and rapid real-time feedback analysis for on-line fluorescent diagnostics of tissues during surgery during surgical operation have been developed [3]. The principle of the new system is based on the use of isolated capillary channels in the volume of the sapphire scalpel for introducing of quartz waveguides, figure 2. One of the waveguides is used for delivering the laser radiation directly to the narrow region of the cutting edge and local excitation of photoluminescence. Another one is used for catching and transfer of photoluminescence to spectrometer. A laser optically coupled directly inside the edge, which acts as an optical wedge, allows obtain effective concentration of the laser energy in the area of the cutting edge of the scalpel for coagulation blood adjacent incision.

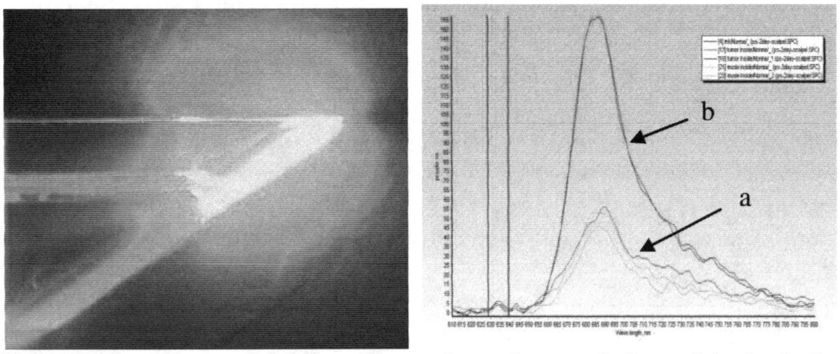

Figure2. Sapphire scalpel for diagnostic and coagulation of incised tissue during a surgical operation (at the left). Fluorescent spectra in different position of scalpel edge: a – normal tissue, b – tumor (at the right).

The system for removal of brain tumors based on a sapphire multi-channel probe with demarcation of borders of a tumor by fluorescent diagnostics with simultaneous coagulation and aspiration was developed, figure 3. It carries out simultaneous laser coagulation for a hemostasis, tumor aspiration via the through channel of a sapphire probe, and also makes local optical measurements of properties of brain tissue for more exact and full removal of malignant tissue.

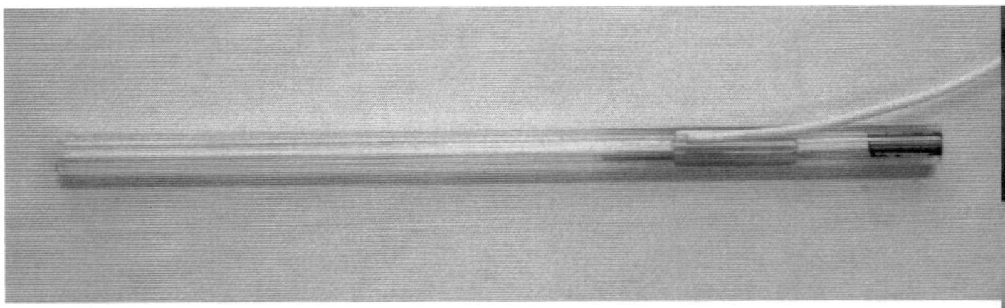

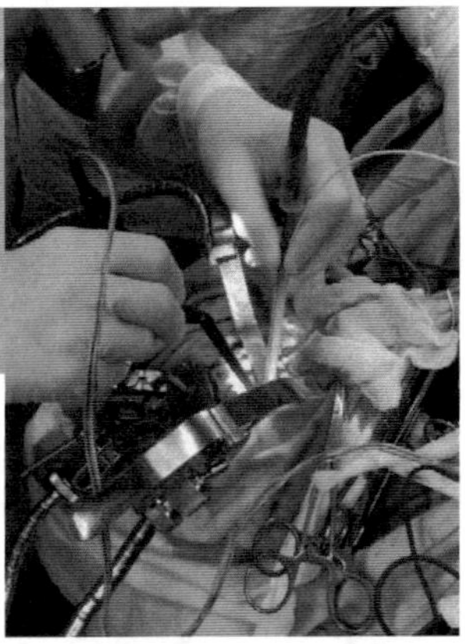

Figure 3. Sapphire multichannel probe for neurosurgery (above). Surgical operation on brain tumor removal using the sapphire probe for neurosurgery (at the right)

The new sapphire cryotips-irradiators made of shaped sapphire are developed. They could be used for cryodestruction of tumors, including big sizes of tumors and allow to carry out simple and convenient sterilization. Usage of the sapphire cryotips may give rise to creation of new techniques of tumors treatment by combination of cryosurgery with laser therapy fluorescent diagnosis due to direct delivery of light radiation to a biological tissue.

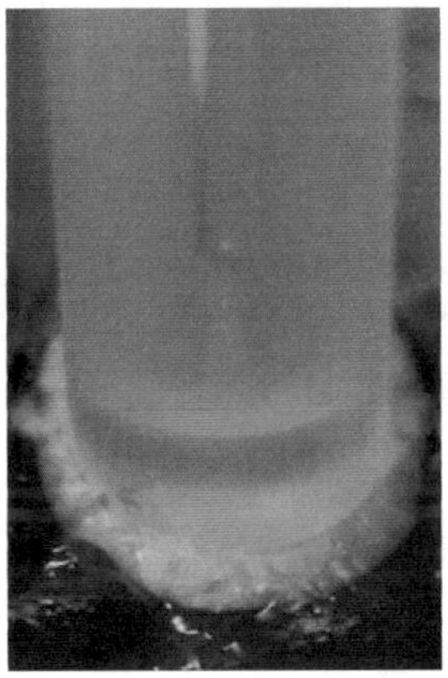

Figure 4. Sapphire cryocooler-lightguide.

Conclusions

The development of sapphire shaping growth methods has given rise to the emergence of essentially new areas of application, including medicine. The application of sapphire can significantly help or completely overcome problems which slow down development of perspective and priority laser methods of treatment in oncology.

Acknowledgements

This work was supported by Russian Foundation for Basic Research (RF) in the frames of the joint research projects RFBR 13-03-01177.

References

[1] Kurlov V N, Rossolenko S N, Abrosimov N V and Lebbou Kh 2010 *Crystal Growth Process Based on Capillarity* ed Th Duffar (John Wiley & Sons) chapter 5 pp 277-354
[2] Shikunova I A, Volkov V V, Kurlov V N and Loshchenov V B 2009 *Bull. Russ. Acad. Sci.: Phys.* **73** 1345
[3] Kurlov V N, Shikunova I A, Ryabova A V and Loschenov V B 2009 *Bull. Russ. Acad. Sci.: Phys.* **73** 1341

Studying the modes of automated destruction of malignant tumors using laser radiation

I V Reshetov[1]*, K G Kudrin[2], D V Davydov[2], V N Kortunov[3], A N Konovalov[3] and G A Varev[4]

[1] I.M.Sechenov First Moscow State Medical University, 119991, Bolshaya Pirogovskaya str. 6, Moscow, Russia
[2] Institute of Improvement of Professional Skill of the Federal Medico-Biological Agency of Russia, 125371, Volokolamskoye shosse 91, Moscow, Russia
[3] Institute on Laser and Information Technologies of the Russian Academy of Sciences, 142092, Pionerskaya str. 2, Troitsk, Moscow Region, Russia
[4] Russian Engineering Club Ltd, 300053, P.O. Box – 712, Tula, Russia

*E-mail: reshetoviv@mail.ru

Abstract. In this work, we have manufactured the experimental setup to study the modes of automated destruction of biological tissue *in vitro*. We have implemented the setup in microsurgical operations, which allows us to verify the modes of tissues destruction.

1. Introduction
Development of robotic systems for surgical applications remains challenging problem of applied physics, optics, and medical science. Robotic surgical systems are minimally invasive and characterized with high speed and precision, which significantly reduces the time of patients stay in medical institutions [1-4]. These advantages of robotic laser surgical systems lead to the increasing manifold of its use in modern robotic surgery [5-8].

One of the advanced laser surgical system is AcuPulse 40 ST (Lumenis Inc., USA). It employs the micromanipulator, and it could be easily adjusted to the operating microscope. This system could perform non-linear surgical incisions, and it is widely applied in ENT, gynaecology, neurosurgery, dermatology, plastic surgery, urology and general surgery. However, despite the listed advantages, the system could not perform automatic tissue evaporation inside specified contour.

In this paper, we study of the modes of automated evaporation of biological tissues, which form the basis for further creation of the robot-assisted laser surgical system.

2. Experimental setup
In order to study the modalities of automated laser evaporation of biological tissue, we have assembled the experimental setup, which consists of

- surgical laser system (SLS) based on CO2 laser operating at 10.6 μm;
- operating microscope Zeiss OPMI MD, with the digital camera connector;
- micromanipulator with the adapter for SLS-connection to a microscope;
- scanning devise for laser beam deflection in the required direction;
- microprocessor-based control system.

The micromanipulator and the system of laser beam deflection are mounted on a surgical microscope, see Figure 1.

Experimental setup could evaporate biological tissue inside required rectangular, triangular, or elliptical contour. The area of the contour could be varied within the field of view of 8 x 8 mm^2, and the duration of tissue evaporation is of about 2 seconds for single puss. The CO2-laser uses in the setup could operate in continuous, pulsed, or superpulsed modes with the wide tenability of output parameters.

3. Studying the modes of the automated tissue destruction

We employ the experimental setup to study different modes of automated destruction of various tissues *in vitro*. By histological examining of the tissues after its exposure with CO2-laser radiation, see figure 2, we determine the optimal modes of the system operation for various surgical tasks.

To verify the destruction modes, we have carried out microsurgical operations, performed on the tissues of the oropharynx and hypopharynx. Namely, 5 operations with the hypopharynx, 10 operations with the larynx, and 40 operations with the skin tissue from the head and the neck have been performed. To improve the accuracy of tissue evaporation, the MAYFIELD skull clamp, fixing the position of the patient head towards the operating table and experimental setup, has been applied. Moreover, while operating the hypopharynx, the rigid endoscope has been applied in addition to the MAYFIELD skull clamp. Thus, the mobility of the surgical field has been significantly reduced.

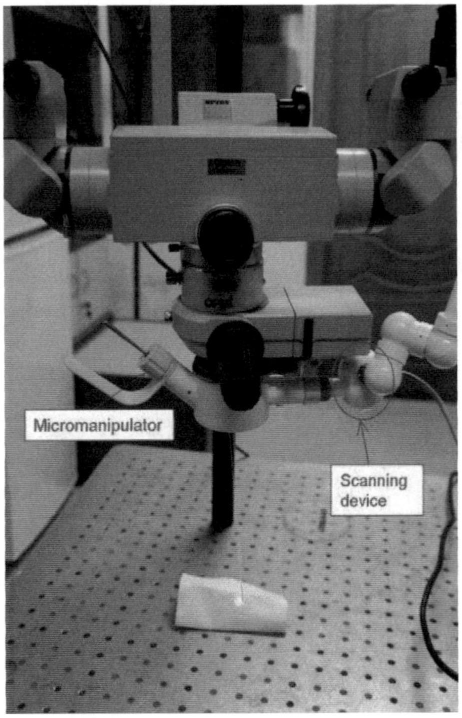

Figure 1. Experimental setup in work.

(a) (b) (c)

Figure 1. Example of laser evaporation of biological tissue:
(a) – initial tissue, (b) – evaporation result, (c) – histological slice.

The experimental results demonstrate, that (i) in equal conditions, the impact of laser radiation on tissue *in vivo* sufficiently differs from the impact on tissue *in vitro*, which originates due the presence of blood flow in tissue *in vivo*; (ii) it is necessary to adjust the laser power to the particular type of tissue to achieve the desired surgical result; (iii) in case of operating the mucosa tissue, the surface should be treated 2 or even 3 time longer in comparison to the skin tissue to achieve equal blood stop effect (this is due to the increased blood supply to the mucosa).

4. Conclusion

In this work, the experimental setup of surgical system for minimally invasive ablation of biological tissues has been developed. This setup has been applied for studying the modes of automated laser destruction of tissues. The regimes of *in vitro* biological tissues evaporation using laser radiation with different parameters have been examined, and the optimal modalities have been determined.

Acknowledgments

The work of I V R., V N K, A N K and G A V was supported by the grant from the Russian Foundation for Basic Research (RFBR), Project # 14-02-12137. K G K. and D V D is thankful for the financial support of the RFBR, Project # 15-29-04894.

References

[1] Gerhardus D 2003 *J. of Healthcare Management* **48** 242–251
[2] Melvin W S, Needleman B J, Krause K R and Ellison E C 2003 *J. of Laparoendoscopic & Advanced Surgical Techniques* **13** 33–36
[3] Talamini M A, Chapman S, Horgan S and Melvin W S 2003 *Surgical Endoscopy* **17** 1521–1524
[4] Ahmed K, Khan M S, Vats A, Nagpal K, Priest O, Patel V, Vecht J A, Ashrafian H et. al. 2009 *International Journal of Surgery* **7** 431–440
[5] Kaufman R and Hibst R 1996 *Lasers Surg. Med.* **19** 324–330
[6] Pozner J N and Goldberg D J 2000 *Dermatol Surg* **26** 733–746
[7] Konovalov A N and Ul'yanov V A 2012 *Applied Optics* **51** 3900-3906
[8] Kortunov V N, Dmitriev A K, Konovalov A N and Ul'yanov V A 2010 *Photodiagnosis and Photodynamic Therapy* **7** 522

Schwinger effect at modern laser facilities

D Blaschke[1,2,3], N T Gevorgyan[4,5], A D Panferov[6] and S A Smolyansky[6]

[1]Institute for Theoretical Physics, University of Wroclaw, 50-204 Wroclaw, Poland
[2]Bogoliubov Laboratory for Theoretical Physics, JINR Dubna, 141980 Dubna, Russia
[3]National Research Nuclear University (MEPhI), 115409 Moscow, Russia
[4]Russian-Armenian (Slavonic) University, 0051 Yerevan, Armenia
[5]Abdus Salam International Centre for Theoretical Physics, I - 34151 Trieste, Italy
[6]Department of Physics, Saratov State University, Saratov, Russia

E-mail: david.blaschke@ift.uni.wroc.pl , gnara@ysu.am , panferovad@info.sgu.ru , smol@sgu.ru

Abstract. The theoretical and experimental investigation of physical processes in strong fields of different nature (electromagnetic, gravitational, etc.) is one of the important directions of modern physics. Particular interest is devoted to the area of extremely strong fields, in which qualitatively new effects become important due to the restructuring of the physical vacuum which accompanies the creation of matter from the vacuum at modern laser facilities. Such kind of time-dependent strong field vacuum effects can be appropriately described within a kinetic theory approach as an effective instrument of theoretical investigations. A short review of recent achievements in the direction of the dynamical Schwinger effect is given in this contribution.

1. Introduction: The physical vacuum in strong, time-dependent fields

The physical vacuum (PV) is understood as the state of a certain region of space, which is characterized by the absence of observable particles and physical fields. In the ordinary sense this is emptiness, a perfect vacuum. Nevertheless, the PV represents a specific material medium. In order to describe its simplest properties, it is necessary to clarify some usual definitions used above in the description of the PV.

Observable particles are understood as localizable material objects characterized by a set of charges (electric, color, ...), momentum $\bar{p}$, mass m, energy ε, etc. It is important that for energy, mass and momentum of a free observable particle the Einstein relation holds

$$\varepsilon = c\sqrt{m^2c^2 + \bar{p}^2} \,. \qquad (1)$$

When this relationship is fulfilled one says that the energy and momentum lie on the mass shell in the four-dimensional space of ε and $\bar{p}$.

Under the absence of physical fields in the PV one understands the vanishing of the mean values of their field strength, which manifests itself in maintaining the state of rest for a test particle under conservation of the corresponding charge and mass.

The PV is one of the fundamental and at the same time complex manifestations of Nature. However, in contrast to the ether of the XIX century, which largely remained a tentative concept, the PV provided answers to the thoughtful questions of the physicists. Direct experimental evidence for the existence of the PV represent such tiny physical effects as the Lamb shift of the energy levels of the hydrogen atom and hydrogen-like atoms, the anomalous magnetic moment of the electron, the Casimir effect (in the simplest case the mutual attraction of uncharged conducting plates).

There are, however, groups of vacuum effects which until now have not found a direct experimental confirmation despite long standing and reliable theoretical predictions. Primarily this concerns the effect of vacuum particle creation under the influence of strong external constant electric fields, predicted in their classical works by F. Sauter (1931) [1], W. Heisenberg and H. Euler (1936) [2] as well as J. Schwinger (1951) who based this effect on a strong foundation in the framework of quantum electrodynamics [3]. Not entirely correct, this effect is often called Schwinger effect (SE). For brevity, we shall use this abbreviation.

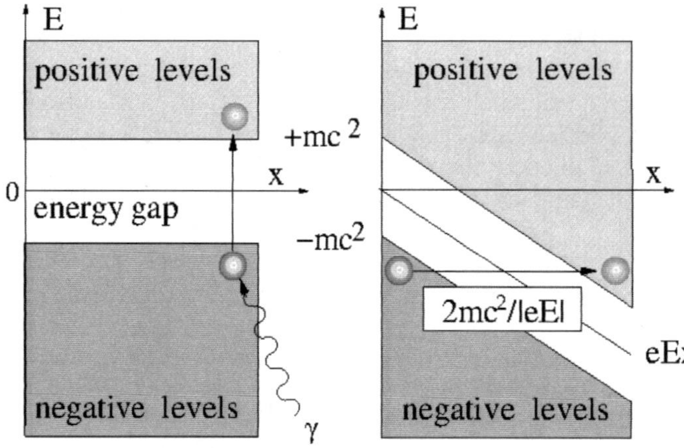

Figure 1. The left panel shows the structure of the energy states of fermions with mass m, in the PV of the classical Dirac theory. The upper region represents allowed energy levels of free states. The lower region the Dirac sea of negative energy states. These regions are separated by an energy gap of width $2mc^2$. The influence of an external electric field is shown on the right panel: the energy gap is converted to an energy barrier accessible for tunneling into the observable region.

The principle of the effect is the following. Under the influence of a static electric field of strength E_0 a particle with charge e a potential energy of eE_0x emerges. This gives rise to a tilt of the energy gap separating the states with positive and negative energies in the Dirac picture of the PV (figure 1, $e < 0$). This is a consequence of the change in the relationship between the energy and momentum (1) under the influence of the field,

$$(\varepsilon - |e|E_0 x)^2 = c^2\left(m^2c^2 + p^2(x)\right). \tag{2}$$

The result is that the electron from the Dirac sea is separated from the region of observable values of the energy by a potential barrier of triangular shape (thick line on the right panel of figure 1), in the inner region of which the momentum $p(x)$ becomes imaginary (forbidden region for classical particle).

From the point of view of quantum mechanics, a particle incident on a potential barrier of arbitrary shape, can penetrate through it with a finite probability (Gamow's tunnel effect [4])

$$w \sim \exp\left\{-\frac{2}{\hbar} i \int_{x_2}^{x_1} dx p(d)\right\} \tag{3}$$

Journal of Physics: Conference Series **672** (2016) 012020 doi:10.1088/1742-6596/672/1/012020

Here we omitted the pre-exponential factor which is usually of the order of unity. Two turning points $x_1 = (\varepsilon - mc^2)/(|e|E_0)$, $x_2 = (\varepsilon + mc^2)/(|e|E_0)$ are found from the condition of vanishing particle momentum $p(x) = 0$. Substituting these values in the formula (3), we obtain

$$w \sim \exp\left\{-\frac{4\mathrm{m}^2c^3}{|e|\hbar E_0}\int_0^1 ds\sqrt{1-s^2}\right\} = \exp\left\{-\frac{\pi m^2c^3}{|e\hbar E_0|}\right\}, \qquad (4)$$

which is in good agreement with the exact solution found by Schwinger for the probability of the formation of electron-positron pairs (EPP) per unit volume per unit time, i.e. for the pair production rate of EPP [3]

$$w = \frac{ce^2E_0^2}{4\pi^3\hbar^2}\exp\left\{\frac{\pi m^2c^3}{|e|\hbar E_0}\right\} = \frac{ce^2E_0^2}{4\pi^3\hbar^2}\exp\left\{-\pi\frac{E_c}{E_0}\right\}, \qquad (5)$$

where E_0 - the constant external field strength, and $E_c = m^2c^3/\hbar|e|$ - the critical field strength. The result (5) is unique in the sense that it can only be obtained using nonperturbative methods, i.e. without the involvement of the standard perturbation theory in the hyperfine structure constant $\alpha = e^2/4\pi\hbar c \simeq 1/137$, with which most of the results of quantum electrodynamics (QED) are obtained. Indeed, the expansion of the exponent in equation (5) for small strengths of the external electric field, i.e. in orders of the ratio $E_0/E_c << 1$, is meaningless.

According to formulas (4) and (5) the effect of vacuum EPP creation should be noticeable if the electric field becomes comparable to the Schwinger critical field strength E_c=1.323·10^{16} V/cm , at which the breakdown of the electromagnetic vacuum occurs. Such gigantic constant fields are unlikely to be achievable even in the long term. For comparison, we give the value of the electric field at the Bohr orbit of the hydrogen atom: $E_B = 10^9$ V/cm.

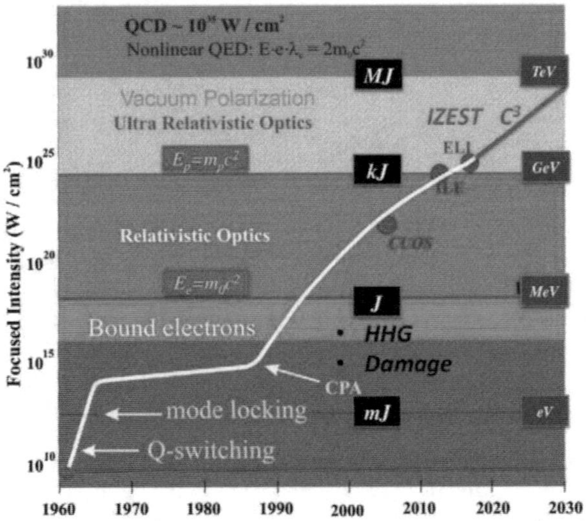

Figure 2. History and growth prospects of the energy density in the focal spot of laser systems (from the work of Mourou et al. [5]).

High hopes for achieving near-critical fields are associated with the rapid development of laser technologies and the increasing power of the generated laser radiation. The idea of using laser light to observe the SE was expressed by F. V. Bunkin and I. I. Tugov already in 1970 [6]. So far, these expectations were not realized. However, existing projects offer hope that the Schwinger limit, estimated by the radiation intensity $I_c = cE_c^2/4\pi$, will be achieved in the near future. Figure 2 shows the growth curve of the optical laser radiation intensity and the expectations for this parameter. Here the most promising optical laser systems are HiPER (Rutherford Laboratory, England), ELI (Extreme Light Infrastructure), XCELS (Russia) and the X-ray Free Electron Laser (XFEL).

The generalization of equation (5) to the case of time-dependent fields is not obvious. This circumstance caused the development of approaches based on different versions of the kinetic theory. It is this direction in which in recent years the majority of results were obtained.

The simplest is the approach proposed by the authors (D.B.B. and S.A.S.) in [7]. Similar to the result by J. Schwinger (5), the kinetic equation (KE) obtained in this work is based on non-perturbative methods of QED and sets from this point of view the standard. It is formulated for a linearly polarized electric field $E(t)$, which depends on time in an arbitrary way, but is spatially homogeneous, i.e. it does not depend on the point of observation. Such conditions are satisfied approximately in the focal spot of two counter propagating laser beams (see figure 3) with dimensions of the order of the laser wavelength.

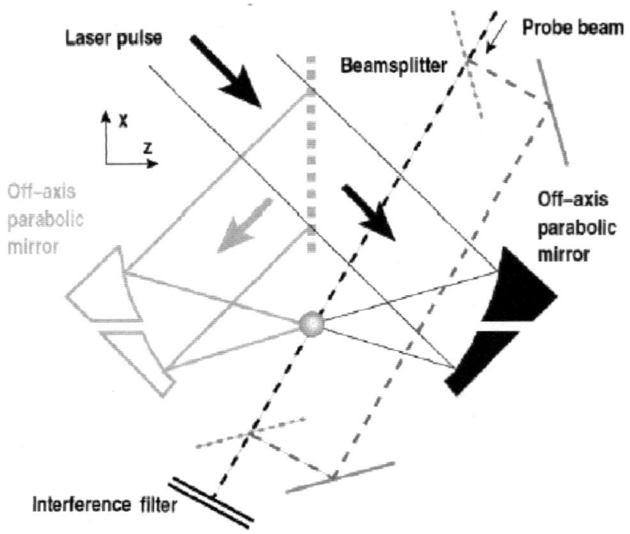

Figure 3. Scheme of a laser collider, providing a standing wave in the focal spot which approximately realizes the situation of a homogeneous, time-dependent and strong electric field ("pump" field, modifying the PV). A probe beam interferometer is sketched which shall measure the refraction index and/or birefringence of the PV region modified by EPP creation due to the "pump" field.

Because of the importance of this KE in the following, we give it in the integro- differential form for the distribution function $f(\vec{p},t)$ of the electrons. If the initial state prior to the application of the external field was a vacuum with zero electric charge, by virtue of charge conservation in the following an equal number of electrons and positrons will be created which move in opposite directions, and the positron component will be described by the distribution function $f(-\vec{p},t)$. The KE describing the vacuum creation of electrons and positrons (we will talk about the birth of an electron-positron pair plasma - EPP) has the form

$$\dot{f}(\vec{p},t)=\frac{1}{2}\lambda(\vec{p},t)\int_{t_0}^{t} dt'\,\lambda(\vec{p},t')\left[1-2f(\vec{p},t')\right]\cos\theta(t,t')\ , \qquad (6)$$

where the vacuum transition amplitude is (here we use units in which $\hbar = c = 1$)

$$\lambda(\vec{p},t)=\frac{eE(t)\varepsilon_\perp}{\varepsilon^2(\vec{p},t)}, \tag{7}$$

$\varepsilon_\perp=\sqrt{m^2+p_\perp^2}$ is the transverse energy, which depends on the components of the momentum $p_\perp$ orthogonal to the direction of the external field $\vec{E}(t)$. Finally,

$$\varepsilon(\vec{p},t)=\sqrt{\varepsilon_\perp^2+(p_{||}-eA(t))^2} \tag{8}$$

is the energy of quasiparticle excitations (electrons or positrons) in the external field, $p_{||}$ - component of the momentum vector parallel to the direction of the external field $\vec{E}(t)=-\dot{\vec{A}}(t)$, where $\vec{A}(t)$ - vector potential of the electromagnetic field. The dot over a symbol denotes here and on the left hand side of the KE (6) the time derivative. Thus the KE (6) describes the strongly nonequilibrium evolution of the EPP by the distribution function $f(\vec{p},t)$. The KE (6) is non-Markovian, i.e. taking into account the whole evolution of the EPP, starting from the moment t_0 when the external field starts: $t_0 \le t' \le t$.

The nontriviality of the KE (6) is largely due to the two-scale evolution of the EPP. The large time scale is determined by the period of the characteristic changes of the external field $T=2\pi/\omega$, where ω is some characteristic frequency. The small time scale corresponds is the Compton time $\tau_c=2\pi/m_e$ comparable with the period of vacuum oscillations in the presence of the field (8). In the absence of an external field, this corresponds to the vacuum fluctuations (Zitterbewegung) with the frequency (1). Usually these scales vary greatly, $T \gg \tau_c$ or $2\pi/T \ll \varepsilon(\vec{p},t)$. The interaction of the time dependent external field with vacuum oscillations leads to complex parametric phenomena accompanying vacuum creation of EPP. These effects can be analyzed by means of numerical and approximate solutions of the KE (6).

Currently, there is a fairly complete picture of the dependence of the structure of the nonequilibrium distribution function on the laser radiation parameters: amplitude, frequency spectrum and laser pulse shape (Section 3). Restricting ourselves to the main characteristics of laser light, the amplitude of the field strength E_0 and the characteristic wavelength λ, the interest is in the following region of parameters:

$$0 \le E_0 \le E_c \ , \ \lambda_c \le \lambda < \lambda_0 \ , \tag{9}$$

where $\lambda_c=2\pi/m$ is the Compton wavelength and $\lambda_0 \sim 1\,\mu\text{m}$ is an upper limit determined by restrictions in the numerical solution of the KE.

However, the distribution function is not accessible to direct experimental observation due to the extreme conditions under which the EPP is born. Observable may be some secondary effects that are available for registration at a sufficient distance from the focal spot. Among these effects are:
- photons radiation from annihilation events in the EPP;
- birefringence of the EPP.

One can expect some other observable manifestations of the EPP created from the PV. Here, the corresponding results have usually only preliminary character. They will be shortly explained in Section 4. To understand the nature of the difficulties that arise here, we first consider the mechanisms of vacuum creation of EPP.

2. Vacuum pair creation process in strong, time-dependent fields

Let us first consider the long-wavelength limit $\lambda \to \infty$ in the sector (9), which corresponds to the constant field case with the field strength E_0. The effect of vacuum particle creation was qualitatively predicted initially based on the Dirac picture of the PV (figure 1), which in the presence of a constant field leads to the tunneling interpretation of the effect. This so-called Schwinger limit (5) is well supported both numerically and analytically by solutions of the KE (6).

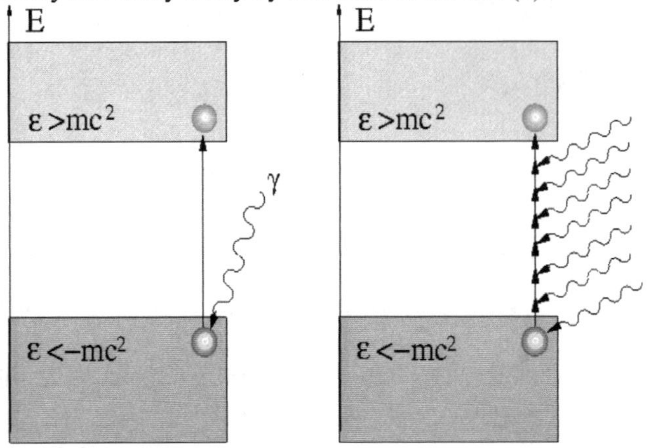

Figure 4. The single-photon (left) and multiphoton (right) mechanisms of the EPP excitation.

In the transition to non-stationary electric fields it is useful representation of the external field as an infinite reservoir of photons with characteristic frequencies determined by the spectral decomposition of the field. Constant field corresponds to the reservoir of photons with an infinitely large wavelength.

We now increase the frequency of the external field up to extremely high frequencies $\nu \geq 2m_e$. Then the energy of one photon of this kind would be sufficient to ensure that there was a breakdown of the energy gap, generating an electron-positron pair. In this speculative limiting case acts the one-photon mechanism [8]! Figure 4 (left) illustrates the effect of such a sort of vacuum photoelectric effect, the photon knocks an electron-positron pair off the PV.

In the transition to lower frequencies (e.g., in the optical or X-ray range) for the breakdown of the energy gap it is necessary that a large number of "soft" photons must act coherently. This corresponds to the multiphoton mechanism of EPP creation. In this case, each of the photons from the external photon reservoir excites the PV to some virtual level (this process is shown on the right panel of Figure 4). From this point of view, the conventional tunneling mechanism can be interpreted as a joint action of an infinite number of coherent photons with frequency tending to zero. It is clear just these multiphoton processes are responsible for the excitation of the EPP in real laser fields.

It is also important to note that during the action of the external field not real commonly observed electrons and positrons are born from the PV, but their quasiparticle analogues having the same charge and spin as the real particles, but with a set of dynamic variables including the characteristics of the external field: a quasimomentum $\vec{P} = \vec{p} - e\vec{A}(t)$ and a quasienergy $\varepsilon(P) = \sqrt{m_e^2 + P^2}$. Here, there is no restriction to linear polarization. Unlike real particles the quasiparticle excitations do not lie on the mass shell (1). It is also important that for the description of processes involving quasiparticles the usual methods such QED are not applicable, such as the S - matrix formalism, which considers the evolution of the system from an infinitely distant past to an infinitely remote future.

Thus, during the laser pulse a quasiparticle EPP is created from the vacuum. After the pulse the electron-positron quasiparticle excitations become real particles and go to the mass shell (1). In this way a residual EPP forms (see figure 5). Both of these stages of evolution give their specific contributions to the formation of secondary observable effects.

Journal of Physics: Conference Series **672** (2016) 012020 doi:10.1088/1742-6596/672/1/012020

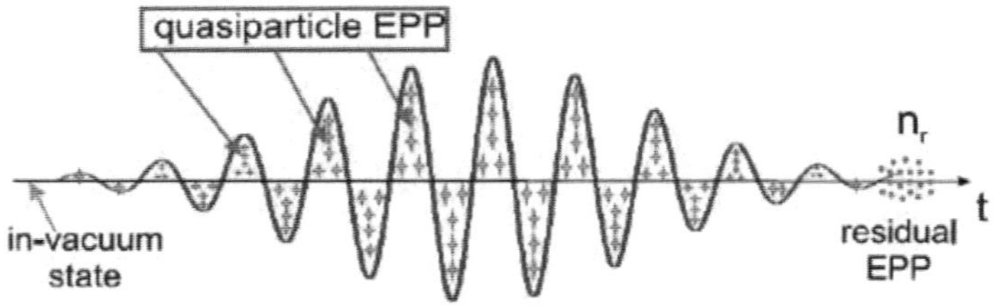

Figure 5. Evolution of the PV under the influence by a pulse of a time dependent electric field.

3. Landscape of EPP distributions

The full information about the EPP in the quasiparticle as well as in the real final state is contained in the distribution function (DF) $f(\vec{p},t)$. Based on the KE (6), below will be considered the reaction of the vacuum on the action of a monochromatic laser pulse with a frequency $\nu(2\pi\nu = \omega)$ and with the Gaussian envelope

$$E(t) = E_0 \cos(\omega t + \varphi) e^{-t^2/2\tau^2}, \tag{10}$$

where τ determines the half-width of the pulse. Typically, $1/\tau << \nu$. Increasing τ, one can model a monochromatic field [9].

The dependence of the DF in the final state on $\vec{p}$ is of particular interest because it determines the spectral characteristics of the experimentally observable real electrons and positrons created by the field. Here appear two characteristic features.

Firstly, an increase of the wavelength in relation to λ_c, corresponding to a reduction of the circular frequency ω in (10) leads to a proportional stretch of the spectrum of produced particles along the direction of the field.

Such a significant complication of the DF is the result of the interference of the contributions of several successive field pulses. Since during the action of each subsequent pulse the quasiparticle density exceeds by many times the residual density accumulated during the preceding stages, the DF saves the memory about the entire history of its evolution.

Increasing the pulse duration by increasing the parameter τ along with the complication of DF form leads to a substantial increase in its maximum value and the total number of produced particles. The total amount of produced particles (per unit volume of the field region) is directly proportional to the pulse duration.

The character of the evolution of DF during the action of the field pulse and its dependence on $\vec{p}$ is usually associated with the adiabaticity parameter $\gamma = E_c\omega / E_0 m$. In figure 6 a general picture of the behavior of this parameter in dependence on the wavelength and the amplitude of the applied field is given [10] (another variant of such kind of landscape is presented in the work [11]).

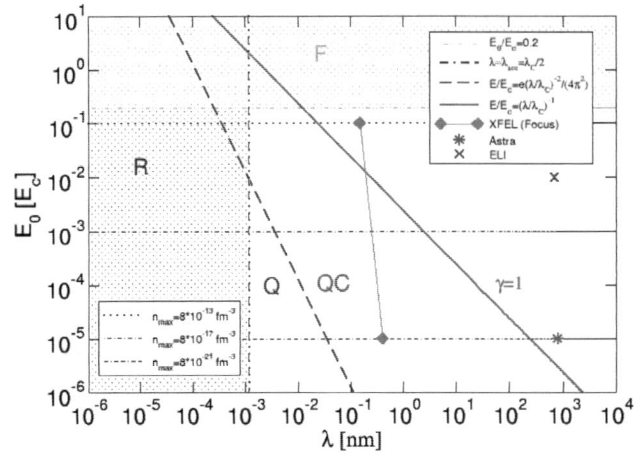

Figure 6. Subdivision of the plane field intensity vs. wavelength in regions of characteristic values for the adiabaticity parameter γ .

The main area is occupied by the "calm valley", where there is no accumulation effect for which the production of EPP would increase with time (i.e. with increasing pulse duration). This accumulation effect of EPP appears in two areas, limiting the "calm valley" on the left and the top. This is a narrow band in the region of extremely short wavelengths $\lambda \leq \lambda_{acc} \approx \pi\lambda_c \approx 0.001$ nm and the region of very strong fields, starting with $E_0 \geq 0.1E_c$. The long-wavelength limit in the accumulation region for strong fields corresponds to the Schwinger production rate of EPP.

4. Experimental verification of the dynamical Schwinger effect at modern laser facilities

Summarizing, we can say that on the basis of numerical and approximate solutions of the KE (6) one can trace the evolution of the DF in sufficient detail during the entire period of the action of the field pulse up to the stage of the freezeout of the EPP. Of course, the limits of applicability of the KE (6) are determined by the model of a linearly polarized spatially uniform external field. These model assumptions can be dropped, however, at the price of a drastic complication of the basic KE. But this does not alter the qualitative picture of the phenomena.

Against this background, it becomes highly actual to perform correct estimations of secondary observable effects such as radiation of annihilation photons from the region of the EPP creation, birefringence etc.

Such a choice of suggested experiments is based on the conditions for the creation of EPP in the focal spot of colliding laser beams, forming there a standing electromagnetic wave (figure 3). The latter is necessary to satisfy the condition of existence of the effect obtained by J. Schwinger. It is clear that in such circumstances the instruments for the detection of EPP generated from the vacuum should be located at sufficient distance from the focal spot.

The above mentioned effects satisfy this requirement. In the case of the annihilation effect there is a finite probability for the annihilation of EPP generated from the vacuum resulting in one or two photons per pulse (figure 7), which can be detected away from the focal spot. Note that single-photon channel being forbidden in the standard QED becomes open in strong fields (i.e. in the quasiparticle stage of evolution EPP).

The idea of the experiment which is based on the effect of birefringence is also simple (see the same figure 3): a test laser beam is split into two, one of which is directed through the region of generation of EPP and afterwards merged again with the other in order to observe their interference pattern. Then, during the term of the main pulse , due to the interaction of one of the rays with the generated EPP its spectral composition is modified which results in changes of the interference pattern.

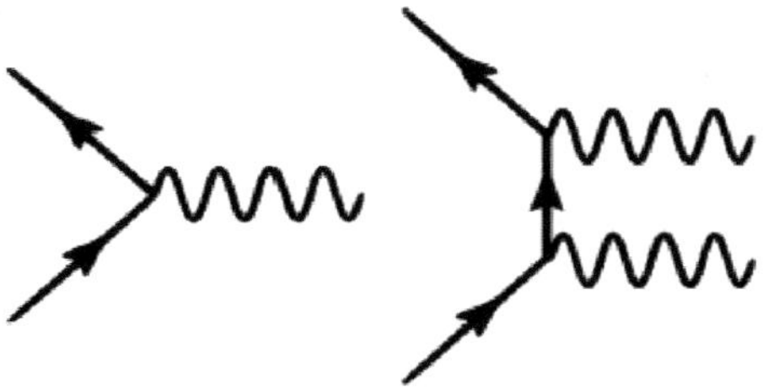

Figure 7. Diagrams of two annihilation channels of EPP produced in the focal spot of two laser beams.

Currently, reliable theoretical estimates of these effects are absent even though quite detailed quantitative descriptions of the effect of EPP creation from the vacuum exist. Preliminary estimates indicate that both of these effects can be detected when the electric field intensity at the level $E_0 \geq 0.1E_c$ will be reached [12]. So in this direction there is still sufficiently elaborated work to be done.

On the other hand, there are doubts as to the possibility of achieving in most advanced laboratory conditions electric fields comparable to the Schwinger limit. The reason is seen in the fact that the Schwinger effect assumes idealized conditions when there is no surrounding matter. In real experiments in a vacuum chamber, covering the area of the focus of the laser beams, there are always trace amounts of gases and vapors which are ionized under the influence of fields which are still very far from the Schwinger limit. The free electrons appearing as a result are accelerated and generate an avalanche of further creation of charged particles. This charged plasma starts screening the electric field and will prevent achieving Schwinger-like field strengths. In this picture, however, also not everything is clear. For example, in the presence of a superstrong field, any electron resulting from ionization, immediately becomes a quasi-particle, indistinguishable from those generated from the vacuum, and the avalanche itself becomes a quasiparticle process.

5. Summary

Despite all contradictions and incompleteness of the current state of the problem, there is no doubt that the vacuum particle creation in strong fields, starting with the creation of matter in the early stages of the Universe, is the basis of a number of phenomena in the laboratory of Nature in all its diverse appearances in atomic and molecular physics, condensed matter physics, astrophysics, physics of ultrarelativistic heavy-ion collisions etc. Thus the key mystery remains unresolved: how proceeds the creation of matter, and what is the basis of the Universe?

Acknowledgements

D.B. acknowledges support by the Polish National Science Center (NCN) within the "Opus" programme grant No. UMO-2014/15/B/ST2/03752. N.T.G. is grateful to the Abdus Salam ICTP, Trieste, Italy for the support provided under the Junior Associateship Programme.

References

[1] Sauter F 1931 *Z. Phys.* **69** 742
[2] Heisenberg W and Euler H 1936 *Z. Phys.* **98** 714
[3] Schwinger J S 1951 *Phys. Rev.* **82** 664
[4] Gamow G 1928 *Z. Phys.* **51** 204

[5] Mourou G A, Tajima T and Bulanov S V 2006 *Rev. Mod. Phys.* **78** 309
[6] Bunkin F V and Tugov I I 1970 *Sov. Phys. Dokl.* **14** 678 ("Possibility of Creating Electron-Positron Pairs in a Vacuum by the Focusing of Laser Radiation")
[7] Schmidt S, Blaschke D, Röpke G, Smolyansky S, Prozorkevich A and Toneev V 1998 *Int. J. Mod. Phys.* E **7** 709
[8] Blaschke D B, Dmitriev V V, Röpke G and Smolyansky S A 2011 *Phys. Rev.* D **84** 085028
[9] Blaschke D, Juchnowski L, Panferov A and Smolyansky S 2015 *Phys. Part. Nucl.* **46** 797
[10] Blaschke D B, Kämpfer B, Schmidt S M, Panferov A D, Prozorkevich A V and Smolyansky S A 2013 *Phys. Rev. D* **88** 045017
[11] Blaschke D, Gevorgyan N, Panferov A D and Smolyansky S A 2014 Proceedings of the Conference "Ewald von Kleist's contribution and its consequences to the development of science", October 11-13, 2013, Kamień Pomorski-Pobierowo, Poland, p. 57-68
[12] Otto A, Seipt D, Blaschke D, Smolyansky S A and Kämpfer B 2015 *Phys. Rev.* D **91** 105018

Experimental study of the light absorption in sea water by thermal lens spectroscopy

A Velásquez[1], E Sira[2], S Silva[2] and H Cabrera[1,3]

[1] Centro Multidisciplinario de Ciencias, Instituto Venezolano de Investigaciones Científicas, Mérida 5101, B. R. Venezuela
[2] Centro de Oceanografía y Estudios Antárticos, Instituto Venezolano de Investigaciones Científicas, Caracas 1020-A, B. R. Venezuela
[3] Associateship Office, International Centre for Theoretical Physics, Trieste 34151, Italy

Email: hcabrera@ictp.it

Abstract. Thermal lens spectroscopy is well known as highly sensitive technique enabling measurements of low absorption and concentration determination of various compounds. The optical absorption coefficients of doubly distilled water and samples of water from different places of the open Ocean and different coastal regions have been measured at 532.8 nm wavelength using this technique. The method enables sensitive, rapid and reproducible determination of small variations of the absorption coefficient which are related with small trace contaminations in sea water.

1. Introduction

Nowadays, the optical properties of liquid water are the basic building blocks on which hydrologic optics is built. To understand the optical properties of the Oceans we must understand the optical properties of the liquid water, which is its major component. The optical properties of both pure liquid water (distilled) and salt waters have been investigated by numerous authors [1-4]. Most early absorption measurements were made by employing conventional absorption spectrophotometer techniques. These techniques require careful evaluation of the optical characteristics of the transmission cell and the preparation of water free from contaminants scattering and absorbing materials. Additionally, due to their low sensitivity require the use of long path cells. The present investigation involves the thermal lens technique that avoids many of the experimental problems associated with the earlier methods [5].

The thermal lens spectroscopy (TLS) is a photothermal technique that has been widely used with high sensitivity and versatility for determination of the absorption coefficient of nearly transparent materials [6-8]. In recent years there has been growing concern about environmental pollution and speciation of trace metals and has received particular attention. Rapid population growth combined with industrialization, urbanization and agricultural intensification has resulted in an increased consumption of natural resources and energy, massive waste production, and steadily growing environmental pollution. The purpose of the present investigation was to measure the absorption coefficient for specimens of sea water from various typical areas of the open Ocean and to analyze different effects of dissolved and suspended materials in the waters of several regions, including

Journal of Physics: Conference Series **672** (2016) 012021 doi:10.1088/1742-6596/672/1/012021

measurements on coastal waters located near to the industries and human populated areas. Furthermore, we compare our absorption values with measurements in doubly distilled water to ascertain in what particulates the sea water differs from distilled water in its optical properties. Information would thus be gained as to whether the salts in solution or any other contaminant exert an appreciable effect on the general transparency, and particularly whether they have any selective action on the absorption coefficient.
A further important object of the present undertaking was to determine to what degree the differences in transparency of various Ocean waters could also be correlated with measurements obtained by other conventional physicochemical methods frequently used in Oceanography.

2. Materials and methods

2.1. Collection of water samples

The locations from which water samples were collected (see Figure 1) were chosen so as to cover the entire range of transparencies or contaminations currently found in the Ocean and coastal areas.

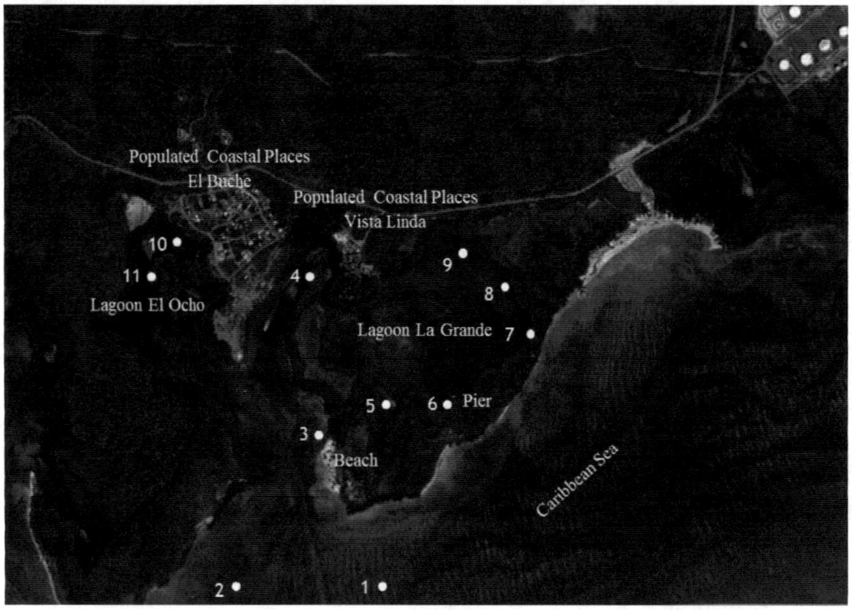

Figure 1. Locations from where water samples were obtained

2.2. Experimental method

The dual-beam thermal lens experimental set-up (see figure 2) is composed of two blocks as follows:

Probe beam and recording block. The He–Ne laser (632.8 *nm*, 10 *mW*, Spectra Physics) generates the probe beam. This beam passes through a 10 *cm* L_1 and 30 *cm* L_2 focal-length lenses, resulting in a 4 *mm* width near collimated probe beam. This beam is reflected by mirrors E_1 and E_2, passes through the dichroic beam combiner *Div.D*, and the sample *M* (contained in a 1 *cm* quartz cuvette). The dichroic mirror *E.D* then reflects the beam towards the 0.5 *mm* pinhole *A* and the photodetector *D* (model DET 110/M, Thorlabs). The output signal of the photodetector is then transformed by the current pre-amplifier *Amp* (model SR570, Stanford Research Systems) and introduced into the digital oscilloscope *Osc* (model TDS 3052, Tektronix), where the signal is digitized, saved, and processed statistically. The radiant flux of the probe beam at the sample surface is 0.5 *mW*.

Excitation beam block. The second harmonic of a diode pumped Neodymium Yttrium Aluminum Garnet (Nd:Yag) CW laser (model GSF32–200, 532.8 *nm*, 200 *mW*, Intelite) delivers the excitation beam through the shutter *T* (model 846, Newport Corporation), the beam splitter *Div D*, and the 180 *mm* focal length lens L_3 to the beam combiner *Div.D*, which reflects it to the sample *M*. The excitation beam passes from the left to the right through the sample cell. The dichroic mirror *E.D* block the excitation beam preventing it from reaching the pinhole *A* and the photodetector. The beam transmitted through the beam splitter *E.D* is sensed by the photodetector *Ref* (model DET 110/M, Thorlabs), which transforms the incident radiant flux into a synchronization signal for the oscilloscope.

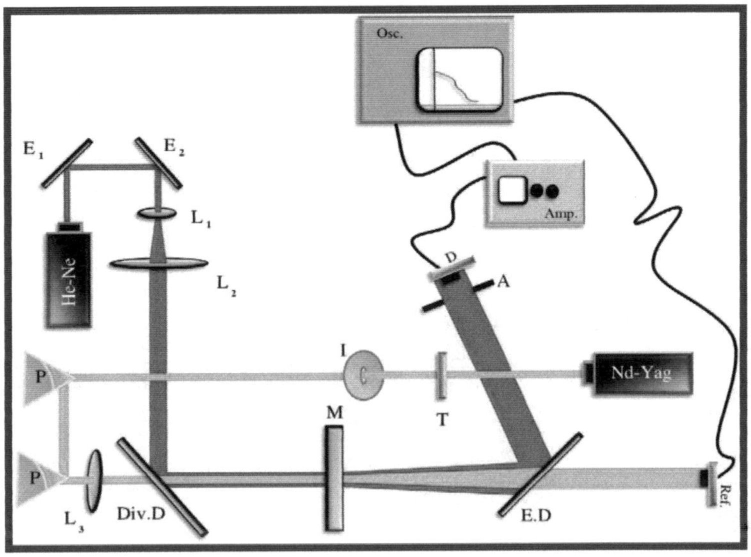

Figure 2. Dual beam TL experimental setup for absorption measurements. E_1, E_2 mirrors; L_1, L_2 lenses; *Div. D* dichroic mirror; *M* sample; *E.D* dichroic mirror; *A* pinhole; *D* photodetector; *Amp* current preamplifier; *Osc* oscilloscope; *T* shutter; *I* beam expander; *P* prism, L_3 lens, *Ref* reference photodetector.

Measurements were taken according to the following procedure: the shutter *T* modulates the beam with a period of 10 *s*, and for each sample the signal was obtained from the average of 128 recordings at the digital oscilloscope. Then the experimental TL signal can be measured as [8]

$$S(z,t)=\frac{T(z,t)-T_0}{T_0} \quad (1)$$

where T_0 is the probe light's transmittance through an aperture with no excitation beam present and $T(z,t)$ is the probe beam's aperture transmittance in the presence of the excitation beam. As we showed before, in a stationary situation, the TL signal is related to the photothermal parameters of the sample by the following expression [8]

$$S(z,t)=\frac{\pi P_e \alpha l}{2\lambda_p k}\frac{dn}{dT} \quad (2)$$

where P_e is the total excitation light power, α, *l, k* and *dn/dT* are the sample's absorption coefficient, thickness, thermal conductivity, and temperature coefficient of the refractive index, respectively. *z* is the sample's position with respect to the waist of the excitation beam radius, *t* is the time and λ_p is the probe beam wavelength. Finally the absorption coefficient can be obtained as

$$\alpha = \frac{S2\lambda_p k}{\pi P_e l dn / dT} \quad (3)$$

Therefore, the measurement methodology can be resumed as follows: the experimental TL signal *S(z,t)* is measured from the experimental data using equation (1), then the corresponding absorption coefficient is calculated using equation (3) with the rest of the photothermal parameters of water very well known, $dn/dT = 0.8\times10^{-4}\,^0C^{-1}$, $k = 5.95\times10^{-3}Wcm^{-1}\,^0C^{-1}$ and fixed experimental values $l = 1cm$, $\lambda_p = 632.8nm$ and P_e=57 *mW*.

3. Results and discussion

3.1. Absorption measurements

In figure 3 the nonprocessed signal obtained at the digital oscilloscope for the sample number 10 is shown. In the figure we indicate the values of the nonperturbed probe light aperture´s transmission T_0 and the probe light´s aperture transmission *T(z,t)* in the presence of the excitation beam. The excitation power used in this experiment is 57 *mW,* the TL signal is then calculated with equation (1) (*S*=-0.017) and the absorption coefficient for this sample was calculated with equation (3) as $\alpha = 8.90\times10^{-4}cm^{-1}$. Additionally, the absorption coefficient for the double distilled water sample was measured as $\alpha = 4.5\times10^{-4}cm^{-1}$.

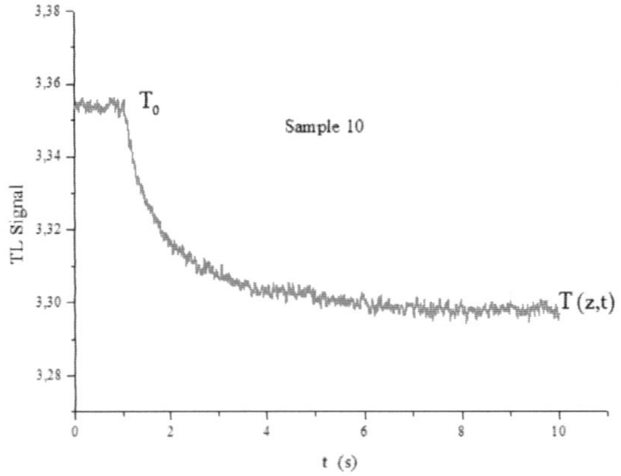

Figure 3. Nonprocessed TL signal for sample number 10 as a function of time for an excitation power of 57 *mW.* The values of the probe beam aperture´s transmission in the absence (T_0) and in the presence *T(z,t)* of the excitation beam are indicated.

We have also performed similar thermal lens measurements for the rest of collected samples and the results are shown in Table 1. Absorption coefficients shown in Table 1 represent the mean value of at least three observations in all cases, what means three samples from the same place and three measurements in each sample.

Table 1. Absorption coefficient values for water samples.

Sample	α (cm^{-1})
1	4.73 x 10^{-4}
2	5.30 x 10^{-4}
3	6.30 x 10^{-4}
4	8.09 x 10^{-4}
5	7.20 x 10^{-4}
6	8.40x 10^{-4}
7	6.80 x 10^{-4}
8	7.35 x 10^{-4}
9	9.25 x 10^{-4}
10	8.90x 10^{-4}
11	8.40 x 10^{-4}

3.2. Physicochemical analysis

Physicochemical parameters of water samples were determined using a multiparametric sound Horiba and the results are shown in Table 2.

Table 2. Physicochemical parameters of water samples.

Sample	**pH**	**Electrical Conductivity ($S\ m^{-1}$)**	**Temperature (°C)**	**Salinity (PSU)**	**Disolved Solids ($g\ L^{-1}$)**
1	8.11	5.63	31.2	37	35
2	8.02	5.5	30.1	36.6	35.3
3	8.07	5.48	29.9	36	35,6
4	7.95	5.6	30.0	37	37
5	7.85	5.67	30.6	38	36
6	7.92	5.57	30.96	37	36
7	7.86	5.73	30.96	38	37
8	7.87	5.74	30.93	38	37
9	7.86	5.73	30.86	38	38
10	7.99	5.46	30.53	36	38
11	8.02	5.52	30.93	37	36

3.3. Discussion

The absorption measurements of the water sample collected from the open Ocean (sample 1) differs little ($\alpha = 4.73\times10^{-4}cm^{-1}$) compared to the absorption value of the doubly distilled water ($\alpha = 4.1\times10^{-4}cm^{-1}$[8] and $\alpha = 4.5\times10^{-4}cm^{-1}$ in the present work); however α increases as one goes from the open Ocean toward the coastal region (see sample 2 and 3), this is due to the higher contamination in this region related with the anthropological activity (note that the sample 3 located near to the beach area presents greater absorption coefficient that the sample 2 which is located far away from the beach.

The absorption of those samples collected close to industrial and populated areas (samples 4, 5, 6, 7, 8, 9, 10, 11) is even higher but the higher value is observed for the sample 9 that was collected in the vicinity of an industrial waste dump. Note that, samples 7 and 8 collected from the same region of the sample 9 have lower value of the absorption. This result may be associated with the fact that the zone where the sample 7 and 8 were collected is very rich in mangrove plants which would also contribute to the purification of some contaminant elements in this region. In general the term mangrove refers to salt-tolerant species of tree or shrub which grow on sheltered shores and in estuaries in the tropics and some sub-tropical regions. They are outstandingly adapted to growing in sea water, which they desalinate by an ultrafiltration process. Mangroves are affected by the freshwater and nutrient supply which they receive from their catchment area, and on the other hand have a strong influence on the adjoining coastal waters and associated ecosystems, therefore we hypothesize that in this zone there is an increased activity of filtering and trapping of water-borne pollutants by mangroves. On the other hand, the sample 6 was not collected from a place located near to any population but presents high absorption because is located near a pier with high contamination activity.

The effect of salinity on absorption must be known to interpret the measured absorption values. Not accounting for the change in the absorption coefficient with salinity could lead to errors in the measurement of absorption by colored dissolved organic matter. Measurements of other optical properties, such as remotely sensed reflectance, which depends on the inherent optical properties of water, must also account for the possible effects of salinity. However, there has been little research done in studying the role of salinity on the optical properties of water at visible wavelengths [9]. The addition of ions to water causes larger, more tightly bound clusters, in which the ions are surrounded by a loose outer shell of water molecules [10]. Therefore, the salinity is expected to affect the molecular structure of water, which in terms affects its absorption properties. Changes in the scattering coefficient with increasing salinity can also influence measurements. However, the reference water used in this study has a salinity of 0 PSU, and sea water samples typically have a salinity of 35 PSU (37 PSU in the present research). In our specific measurements (see Table 2) the salinity is not significantly different from sample to sample and on the other hand the electrical conductivity, that depends on the ionic species and dissolved solids, correlated well with salinity having similar tendencies. Additionally, the absorption by particulate material or scattering absorption was also neglected because the TLM method is free of scattering. Consequently, samples collected from the open Ocean (1, 2, 3) were affected by the absorption of dissolved organic materials and these collected from the coastal zone (4, 5, 6, 7, 8, 9, 10, 11) by the organic materials but also by the waste residues of industries with additional metals and inorganic traces. Generally, the pH of open Ocean waters oscillates between 8.1 and 8.3, taking their lowest values near to the coastal areas indicating some degree of photosynthesis or organic contamination. These tendencies are similar in our measurements in which the sample of the zone1 has a pH 8.11 and for coastal samples the pH decreases.

Finally, TLS is a non-specific detection technique and therefore requires prior separation of analytes species for a quantitative analysis such as metal speciation. Metal speciation refers to the identification and quantification of chelated, organometallic or free metal ions or is very important to efficiently separate their oxidation states in a particular sample. Ion chromatography (IC) [11] has become one of the most widely used analytical method for the speciation of these elements due to its potentialities. The main advantages of IC are the capability for speciation and, furthermore, on-line matrix elimination and sample preconcentration is possible using an appropriate column, which can lower the detection limits. One of the detection methods to improve the detection limits in IC is TLS. The combination of TLS and IC is a good option for quantitatively account the degree of contamination in natural and sea waters.

4. Conclusions

In this paper, a TLS method was applied for measuring the absorption in water samples obtained from different places of the open Ocean and the coastal region of the Caribbean Sea to qualitatively account for a degree of contamination due to the anthropological action of humans. Collectively, our results suggest that there is an increased degree of contamination in the coastal zone with high industrial and human activity. The accuracy of the applied method does not depends on the water salinity indicating that it is possible to account for the influence of organic, inorganic or heavy metal species. This work confirms the applicability of TLS qualitatively detection of trace contaminants in water samples and the possibility of quantitative determination if combined with other separation methods.

Acknowledgments

We thank IVIC for the financial support received during this research.

References

[1] Jerlov N G 1968 *Optical Oceanography* (New York: Elsevier)
[2] Jerlov N G 1976 *Marine Optics* (New York: Elsevier)
[3] Hale G M and Querry M R 1973 *Appl. Opt.* **12** 555
[4] Tam A C and Patel C K N 1979 *Appl. Opt.* **18** 3348
[5] Cabrera H, Cordido F, Velásquez A, Moreno P, Sira E and López-Rivera S A 2013 *C. R. Mecanique.* **341** 372
[6] Cabrera H, Korte D and M Franko 2015 *Rev. Sci. Instrum.* **86** 053701
[7] Cabrera H, Korte D and M Franko 2015 *Int J. Thermophys.* **36** 2434
[8] Marcano A, Cabrera H, Guerra M, Cruz R A, Jacinto C and Catunda T 2006 *J. Opt. Soc. Am. B* **23** 1408
[9] Ravisankar M, Raghunath A T, Sathianandan K and Nampori V P N 1988 *Appl. Opt.* **27** 3887
[10] Dera J 1992 *Marine Physics* (Elsevier, Amsterdam)
[11] Small H, Stevens T S and Bauman W S 1975 *Anal. Chem.* **47** 1801
[12] Sikovec M, Franko M, Novic M and Veber M 2001 *J. Chromatogr. A* **920** 119